Name ______________________ Class ________

Skills Worksheet

Directed Reading

Section: Global Change

In the space provided, write the letter of the term or phrase that best completes each statement or best answers each question.

______ **1.** Scientists have discovered that acid rain is caused by
- **a.** acid evaporating into the atmosphere.
- **b.** sulfur introduced into the atmosphere from burning coal.
- **c.** hydrochloric acid in clouds.
- **d.** gas introduced into the atmosphere from animal wastes.

______ **2.** Which of the following statements about acid rain is NOT true?
- **a.** Rain and snow in the northeastern United States are more acidic than in the rest of the United States.
- **b.** In the United States and Canada, many lakes are "dying" as their pH levels fall below 5.0.
- **c.** Acid rain is causing damage to forests.
- **d.** Tall smokestacks prevent smoke from polluting the atmosphere.

______ **3.** A hole in the ozone layer has formed over
- **a.** Antarctica.
- **b.** North America.
- **c.** Australia.
- **d.** None of the above

______ **4.** The primary cause of the destruction of ozone is
- **a.** a higher concentration of sulfuric acid in acid precipitation.
- **b.** higher carbon dioxide levels in the atmosphere.
- **c.** chlorofluorocarbons (CFCs) reaching the upper atmosphere.
- **d.** increased amounts of ultraviolet radiation reaching our planet.

______ **5.** Exposure to high levels of ultraviolet radiation can lead to
- **a.** more productive food crops.
- **b.** larger animal populations.
- **c.** warmer summers and cooler winters.
- **d.** skin cancer and cataracts.

______ **6.** The protective shield of ozone is needed because ozone
- **a.** is part of the nitrogen cycle.
- **b.** absorbs solar wind.
- **c.** absorbs ultraviolet radiation.
- **d.** is a source of oxygen.

______ **7.** The use of CFCs has been banned or limited in many countries because
a. CFCs are carcinogens.
b. Ultraviolet light breaks down CFCs and the resulting chlorine destroys ozone.
c. CFCs cause asthma.
d. less expensive coolants can be used in refrigerators and air conditioners.

______ **8.** All of the following are greenhouse gases EXCEPT
a. chlorine.
b. carbon dioxide.
c. methane.
d. nitrous oxide.

______ **9.** The chemical bonds in carbon dioxide molecules that absorb solar energy as heat radiates from Earth is known as
a. the ozone layer.
b. the greenhouse effect.
c. acid rain.
d. CFCs.

Name ______________________ Class ______________ Date ____________

Skills Worksheet

Directed Reading

Section: Effects on Ecosystems

Read each question, and write your answer in the space provided.

1. What is biological magnification?

____

____

____

____

2. What are agricultural chemicals?

____

____

3. What happened to the oil tanker in 1989, and what impact did it have on the environment?

____

____

4. Why did Lake Erie become polluted?

____

____

5. What effect does DDT have on birds?

____

____

6. Why is the rosy periwinkle an important species?

____

____

7. Why is the rosy periwinkle in danger of being wiped out?

____

____

8. Why are people concerned about the destruction of tropical rain forests?

____

____

____

9. Why is the loss of topsoil a great concern?

____

____

10. What are aquifers?

____

____

11. How is our ground water being wasted?

____

____

____

In the space provided, write the letter of the term or phrase that best completes each statement or best answers each question.

______**12.** The rate of population growth has increased since 1650 because of which of the following?
- **a.** an increase in the birth rate
- **b.** a decrease in the birth rate
- **c.** an increase in the death rate
- **d.** a decrease in the death rate

______**13.** The United Nations predicts that the world population will stabilize at
- **a.** 6 billion.
- **b.** 6.8 billion.
- **c.** 8.5 billion.
- **d.** 9.7 billion.

Name ______________________ Class ______________ Date ____________

Skills Worksheet

Directed Reading

Section: Solving Environmental Problems

Complete each statement by underlining the correct term or phrase in the brackets.

1. In the early 1990s, there was an international agreement to stop [CFC / DDT] production.

2. All cars in the United States are required by law to have [filters / catalytic converters].

3. The Clean Air Act of 1990 requires that power plants install [scrubbers / recycling devices] in their smokestacks to reduce pollution.

4. The gasoline [tax / act] can reduce the use of gasoline.

The following are five components to solving an environmental problem successfully. In the space provided, describe what each component entails.

5. assessment

____

____

6. risk analysis

____

____

7. public education

____

____

8. political action

____

____

9. follow-through

____

____

Name ______________________ Class ______________ Date ____________

Skills Worksheet

Active Reading

Section: Global Change

Read the passage below. Then answer the questions that follow.

Acid rain forms when coal-burning power plants send smoke high into the atmosphere through smokestacks that are often more than 300 m (969 ft) tall. This smoke contains high concentrations of sulfur because the coal that plants burn is rich in sulfur. The intent of those who designed the power plants was to release the sulfur-rich smoke high into the atmosphere, where winds would disperse and dilute it.

Scientists have discovered that the sulfur introduced into the atmosphere by smokestacks combines with water vapor to produce sulfuric acid. Rain and snow carry the sulfuric acid back to Earth's surface. This acidified precipitation is **acid rain.**

SKILL: RECOGNIZING CAUSE AND EFFECT

Read each question, and write your answer in the space provided.

1. What are the cause and effect of the relationship described in the second sentence of this passage?

____

____

2. Why did power plant designers build tall smokestacks?

____

____

3. What actually occurs when the sulfur-rich smoke is released into the atmosphere?

____

____

4. What causes the sulfuric acid to reach Earth's surface?

____

____

SKILL: SEQUENCING INFORMATION

Study the following steps in the formation of acid rain. Determine the order in which the steps take place. Write the number of each step in the space provided.

______ **5.** Sulfur-rich smoke is released into the atmosphere.

______ **6.** Acidified precipitation falls to Earth.

______ **7.** Smoke containing high concentrations of sulfur is produced.

______ **8.** Sulfur combines with water vapor to produce sulfuric acid.

______ **9.** Coal is burned in power plants.

In the space provided, write the letter of the phrase that best completes the statement.

______ **10.** Acid rain is an effect of
- **a.** wind currents.
- **b.** burning of coal.
- **c.** evaporation of water.
- **d.** Both (a) and (b)

Name ______________________ Class ______________ Date ______________

Skills Worksheet

Active Reading

Section: Effects on Ecosystems

Read the passage below. Then answer the questions that follow.

The world's population exceeded 6 billion in early 1999, and the annual increase is now about 94 million people. About 260,000 people are added to the world's population each day, or about 180 people every minute. Population growth is fastest in the developing countries of Asia, Africa, and Latin America. Growth is slowest in the industrialized countries of North America, Europe, Japan, New Zealand, and Australia. The population growth in the United States is only 0.8 percent, less than half of the global rate. Most European countries are growing even more slowly, and the populations of Germany and Russia are actually declining. In contrast, as of 1996, Nigeria's population was increasing by about 3.05 percent per year.

SKILL: READING EFFECTIVELY

Read each question, and write your answer in the space provided.

1. What does the word *about* in the first sentence have to do with the world's annual rate of population growth?

____

____

2. About how many people are added to the world's population during one of your biology classes?

____

____

3. Where is population growth the fastest?

____

____

4. Where is population growth the slowest?

____

____

5. Based on the passage, how could you determine the world's population growth rate?

____

____

____

In the space provided, write the letter of the phrase that best completes the statement.

______ **6.** The population growth rate in most European countries is
- **a.** less than 0.8 percent.
- **b.** about 0.8 percent.
- **c.** more than 1.6 percent.
- **d.** more than 3.05 percent.

Name ______________________ Class ______________ Date ____________

Skills Worksheet

Active Reading

Section: Solving Environmental Problems

Read the passage below. Then answer the questions that follow.

There are five components to successfully solving any environmental problem.

Assessment: The first stage is scientific analysis of the problem, the gathering of information about what is happening. To construct a model of the ecosystem, data must be collected and experiments must be performed. A model makes it possible to describe how the ecosystem is responding to the situation. It is then used to make predictions about the future course of the ecosystem.

Risk analysis: Using the information obtained by scientific analysis, it is possible to predict the consequences of environmental intervention. It is also essential to evaluate any negative effects associated with a plan of action.

Public education: When it is possible to describe alternative courses of action, the public must be informed. This involves explaining the problem in understandable terms, presenting the alternative actions available, and explaining the probable costs and results of the different choices.

Political action: The public, through its elected officials, selects and implements a course of action. Individuals can be influential at this stage by exercising their right to vote and by contacting their elected officials.

Follow-through: The results of any action should be carefully monitored to see if the environmental problem is being solved.

SKILL: READING EFFECTIVELY

Read each question, and write your answer in the space provided.

1. What type of information is gathered during a scientific analysis of the problem?

____

____

2. How does a model of the ecosystem help assess the problem?

____

____

____

3. What three actions are involved in public education?

____

____

____

4. How can individuals influence selection of a specific course of action?

____

____

SKILL: SEQUENCING INFORMATION

Study the following steps that show the sequence of actions taken to successfully solve an environmental problem. Determine the order in which the steps take place. Write the number of each step in the space provided.

______ **5.** Actions are taken to resolve the problem.

______ **6.** A model of the ecosystem is made.

______ **7.** The ecosystem is monitored to see if the action has solved the problem.

______ **8.** Information about the problem and possible action is presented to the public.

______ **9.** Data is collected, and experiments are conducted.

______ **10.** Predictions are made about the consequences of environmental intervention.

______ **11.** Based on a model, predictions are made about the future course of the ecosystem.

In the space provided, write the letter of the term or phrase that best completes the statement.

______ **12.** A model of an ecosystem is based on
- **a.** data collected from the ecosystem
- **b.** experiments.
- **c.** public opinion.
- **d.** Both (a) and (b)

Name ______________________________ Class ______________ Date ____________

Skills Worksheet

Vocabulary Review

Complete each statement by writing the correct term or phrase in the space provided.

1. When sulfur combines with water vapor to form sulfuric acid, the resulting precipitation is called ____________________ ____________________ .

2. The major cause of ozone destruction is a class of chemicals, invented in the 1920s, called ____________________ .

3. The warming of the atmosphere that results from greenhouse gases is known as the ____________________ ____________________ .

4. As molecules of chlorinated hydrocarbons pass up through the trophic levels of the food chain, they become increasingly concentrated. This process is called ____________________ ____________________ .

5. Porous rock reservoirs for ground water are called ____________________ .

In the space provided, write the letter of the description that best matches the term or phrase.

______ **6.** global change

______ **7.** ozone hole

______ **8.** malignant melanoma

______ **9.** greenhouse gases

______ **10.** chlorinated hydrocarbons

______ **11.** carcinogen

a. a class of compounds that includes DDT, chlordane, lindane, and dieldrin

b. examples include acid rain and ozone destruction

c. a zone in the atmosphere with a below-normal concentration of ozone

d. a potentially lethal form of skin cancer

e. a cancer-causing agent

f. gases with insulating effects

Skills Worksheet

Science Skills

Interpreting Maps

Use the map below, that shows the distribution of acid rainfall in the United States, to answer the following questions.

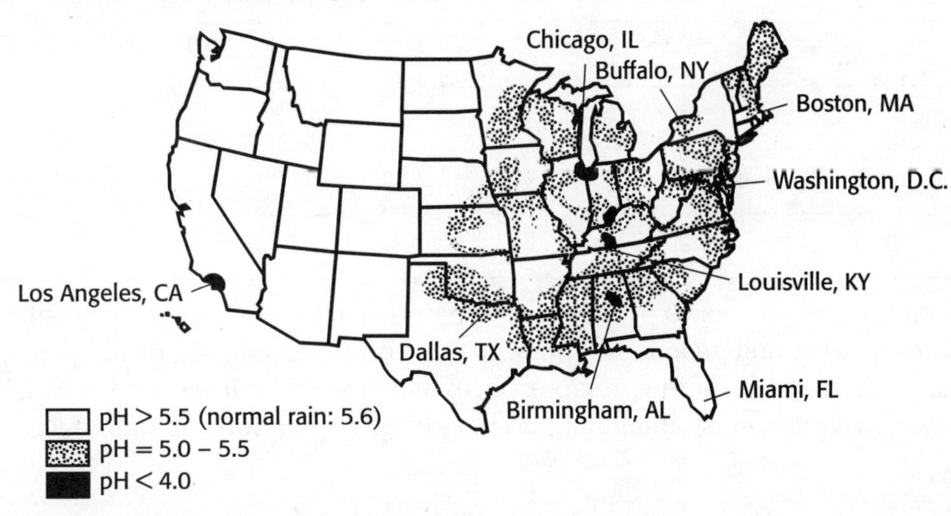

Read each question, and write your answer in the space provided.

1. How is acid rain formed?

____

____

____

____

2. What areas shown in the map above experience the most acidic rain?

____

____

____

____

3. In what region of the United States does most acid rain fall? What conclusions can you draw from this fact?

____

____

____

____

4. Weather patterns in North America generally move from west to east. Where do you think most of the chemicals that cause acid rain are generated? Explain.

____

____

____

____

____

____

5. Normal rainwater has a pH of approximately 5.6. In lakes and ponds with a pH of 5, salamanders and other amphibians often develop deformities or die. Based on the map on the previous page, which section of the United States should have the most abundant and healthiest populations of amphibians?

____

____

Name ______________________ Class ______________ Date ______________

Skills Worksheet

Concept Mapping

Using the terms and phrases provided below, complete the concept map showing the impact of humans on the environment.

acid rain	global change	pollution
agricultural chemicals	ground water	topsoil
ecosystems	ozone depletion	

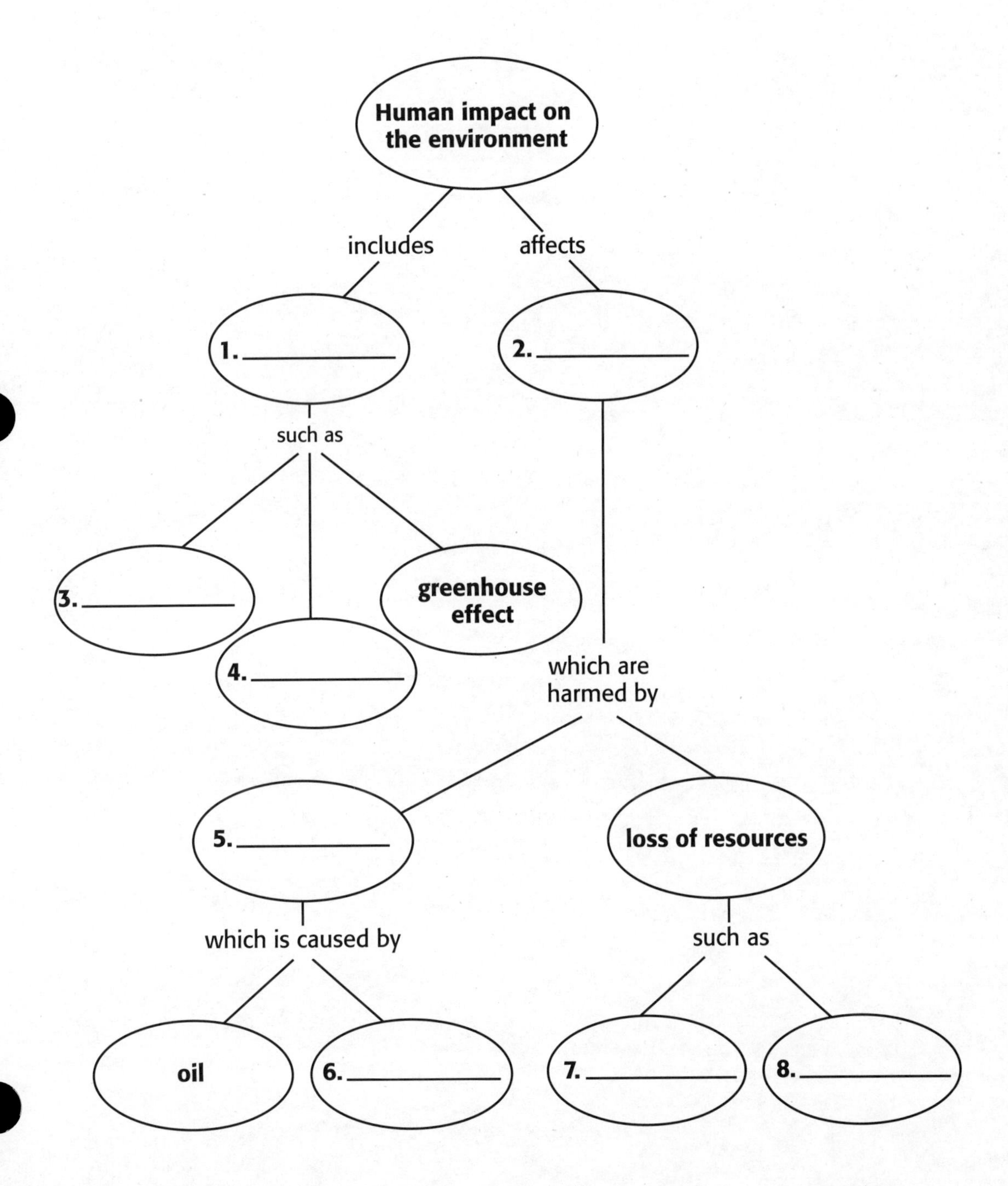

Name ______________________ Class ______________ Date ____________

Skills Worksheet

Critical Thinking

Work-Alikes

In the space provided, write the letter of the term or phrase that best describes how each numbered item functions.

_____ **1.** acid rain

_____ **2.** carbon dioxide buildup in the atmosphere

_____ **3.** aquifer

_____ **4.** human population growth

a. chain reaction

b. boomerang

c. underground water channel

d. greenhouse

Cause and Effect

In the space provided, write the letter of the term or phrase that best matches each cause or effect given below.

	Cause	Effect
5.	_______________	increased global temperatures
6.	increased UV radiation	_______________
7.	_______________	sulfur dioxide, carbon dioxide, and soot emissions cut by 30% in 10 years
8.	oil tanker loaded with oil above water line	_______________

a. global increase in efforts to reduce pollution

b. 1,600 km of Alaska polluted and thousands of marine animals killed in 1989

c. increased atmospheric carbon dioxide for last 50 years

d. skin cancer, cataracts, and cancer of the retina

Trade-offs

In the space provided, write the letter of the bad news item that best matches each numbered good news item below.

Good News

_____ **9.** Humans rely on Earth for all food and materials.

_____ **10.** Laws were passed in 1972 to ban the use of DDT in the United States.

_____ **11.** Gasoline has been taxed to reduce pollution.

_____ **12.** The global rate of population growth is declining.

_____ **13.** Much energy is obtained by burning coal.

Bad News

a. The smoke produces acid rain.

b. Developing countries continue to use it.

c. It is an artificial price hike imposed by the government.

d. There may be none left for future generations.

e. The world population will increase to 8.5 billion by 2025.

Linkages

In the spaces provided, write the letters of the two terms or phrases that are linked together by the term or phrase in the middle. The choices can be placed in any order.

14. ______ sulfuric acid produced ______

15. ______ ozone is changed to oxygen ______

16. ______ carbon dioxide build-up in the atmosphere ______

17. ______ overgrazing and poor land management ______

18. ______ your participation in the solution ______

a. acid rain
b. your knowledge of environmental problems
c. sulfur from smoke joins with water vapor
d. topsoil accumulates as decayed remains of plants and animals
e. your contribution to the solution of environmental problems
f. greenhouse effect
g. CFCs are released into atmosphere
h. burning fossil fuels releases carbon dioxide
i. one-fourth of topsoil lost since 1950
j. ozone layer holes over South Pole

Analogies

An analogy is a relationship between two pairs of terms or phrases written as a : b :: c : d. The symbol : is read as "is to," and the symbol :: is read as "as." In the space provided, write the letter of the pair of terms or phrases that best completes the analogy shown.

______**19.** ozone hole : CFCs released into atmosphere ::
a. acid rain : carbon dioxide
b. sulfur : ozone
c. increased UV light : presence of ozone hole
d. decreased ozone : sulfur

______**20.** assessment : risk analysis ::
a. risk analysis : assessment
b. public education : risk analysis
c. political action : assessment
d. public education : political action

______**21.** oil spillage at sea : major losses of marine animals ::
a. loss of topsoil in the United States : rain forest destruction
b. aquifer pollution : loss of marine animals
c. poor disposal of chemical wastes : ground-water pollution
d. rain forests cut for timber : poor disposal of chemical wastes

______**22.** milk : replaceable ::
a. lumber : nonreplaceable
b. ground water : replaceable
c. original topsoil : replaceable
d. species of plant or animal : nonreplaceable

Name ______________________ Class ______________ Date ______________

Assessment

Test Prep Pretest

In the space provided, write the letter of the term or phrase that best completes each statement or best answers each question.

_____ **1.** When the sulfur in the atmosphere combines with water vapor, the result is
a. ozone.
b. CFCs.
c. acid rain.
d. ultraviolet radiation.

_____ **2.** Global levels of carbon dioxide are
a. rising.
b. remaining constant.
c. falling.
d. too low to be measured accurately.

_____ **3.** All of the following are considered nonreplaceable resources EXCEPT
a. topsoil.
b. wood.
c. ground water.
d. animal and plant species.

_____ **4.** If current birth rates and death rates remain constant, the world's population will double in
a. 20 years.
b. 30 years.
c. 40 years.
d. 60 years.

_____ **5.** In which of the following countries is population growth most rapid?
a. United States
b. Nigeria
c. Australia
d. Japan

_____ **6.** Worldwide efforts to reduce pollution include all of the following EXCEPT
a. severe restrictions on the use of DDT.
b. taxation and legislation.
c. international agreements to stop CFC production.
d. an international agreement to close all coal-burning facilities.

_____ **7.** The first stage of addressing an environmental problem is
a. assessment.
b. risk analysis.
c. public education.
d. political action.

Complete each statement by writing the correct term or phrase in the space provided.

8. Because of the current condition of the ozone layer, more ______________ ______________ is reaching Earth's surface.

9. The insulating effect of various gases in Earth's atmosphere is known as the ______________ ______________.

10. The increase in global temperatures is called ______________ ______________.

11. Examples of chemical pollutants released into the global ecosystem by the agriculture industry are ______________ ______________, and ______________.

12. Since 1650, the human ______________ ______________ has remained constant, and the ______________ ______________ has fallen steadily.

13. In the United States, the population ______________ ______________ is less than half the global rate.

14. Washing cars and watering lawns less often and using efficient faucets are ways to conserve ______________ ______________.

15. The Clean Air Act of 1990 requires that power plants install ______________ on their smokestacks.

Read each question, and write your answer in the space provided.

16. How does the presence of the ozone layer affect life on Earth?

____

____

____

17. Explain the relationship between the greenhouse effect and global warming.

____

____

____

____

____

18. Explain biological magnification.

____

____

____

____

____

19. What has happened to the human death rate in the past several hundred years? Explain.

____

____

____

____

____

20. List the five components necessary to solve an environmental problem successfully.

____

____

Name ______________________________ Class ______________ Date ____________

Assessment

Quiz

Section: Global Change

In the space provided, write the letter of the description that best matches the term or phrase.

_____ **1.** acid rain

_____ **2.** global warming

_____ **3.** CFCs

_____ **4.** ultraviolet radiation

_____ **5.** ozone

_____ **6.** greenhouse effect

a. class of chemicals used as heat exchangers that can break down and destroy ozone

b. the absorption of solar energy by insulating gases such as carbon dioxide

c. creates a protective shield in the upper atmosphere

d. a period of more than a century with steadily increasing average temperatures worldwide

e. exposure to this causes increased occurrences of some diseases, including cancer

f. precipitation high in sulfur

In the space provided, write the letter of the term or phrase that best completes each statement or best answers each question.

_____ **7.** In the northeastern United States, rain and snow have an average pH of which of the following?

a. 3.5–4.0
b. 4.0–4.5
c. 5.5–6.0
d. 7.0–7.5

_____ **8.** CFCs were commonly used as

a. coolants in refrigerators.
b. foaming agents in plastic-foam cups
c. aerosol propellants in spray cans.
d. All of the above

_____ **9.** Greenhouse gases include

a. carbon monoxide.
b. methane.
c. oxygen.
d. All of the above

_____ **10.** Evidence of the disintegration of Earth's ozone shield is found as far back as

a. 1960.
b. 1978.
c. 1980.
d. 1985.

Name ______________________ Class ______________ Date ____________

Assessment

Quiz

Section: Effects on Ecosystems

In the space provided, write the letter of the description that best matches the term or phrase.

______ **1.** aquifer

______ **2.** biological magnification

______ **3.** rain forest

______ **4.** rosy periwinkle

______ **5.** agricultural chemicals

a. a flower from which anticancer drugs have been developed

b. porous rock reservoir where ground water is stored

c. pollutants that include fertilizers, herbicides, and pesticides

d. concentration of chemicals as they progress through the food chain

e. disappearing habitat that is rich in species

In the space provided, write the letter of the term or phrase that best completes each statement or best answers each question.

______ **6.** Examples of nonrenewable resources that are being consumed or destroyed include all of the following EXCEPT
a. topsoil
b. ground water
c. chlorofluorocarbons
d. species of living things

______ **7.** A pesticide that is now banned, but previously caused problems with predatory bird eggs during reproduction is
a. chlordane.
b. lindane.
c. DDT.
d. dieldrin.

______ **8.** Rainforests are being destroyed for
a. farmland.
b. cattle ranching.
c. timber.
d. All of the above

______ **9.** The proportion of well-known species that are currently on the brink of extinction is approximately which of the following?
a. 10 percent
b. 15 percent
c. 25 percent
d. 40 percent

______ **10.** Topsoil is removed and lost through
a. turning over the soil to eliminate weeds.
b. allowing animals to overgraze.
c. practicing poor land management.
d. All of the above

Name ______________________________ Class ______________ Date ____________

Assessment

Quiz

Section: Solving Environmental Problems

In the space provided, write the letter of the description that best matches the term or phrase.

_____ **1.** assessment

_____ **2.** follow-through

_____ **3.** political action

_____ **4.** public education

_____ **5.** risk analysis

a. gathering information about the problem

b. the public selects and implements a plan of action

c. scientists predict the consequences of different types of intervention

d. the results of any action are monitored to see if the environmental problem is being solved

e. explaining the problem in understandable terms and describing alternative actions

In the space provided, write the letter of the term or phrase that best completes each statement or best answers each question.

_____ **6.** The United States has successfully reduced some pollution by restricting the use of

a. DDT.
b. asbestos.
c. dioxin.
d. All of the above

_____ **7.** Cars reduce the pollution they release by means of which of the following?

a. scrubber
b. catalytic converter
c. gas-powered engine
d. All of the above

_____ **8.** The purpose of adding a tax to pollution-causing activities is

a. to make them more expensive and lower the amount of activity.
b. to exploit the activity in order to raise money.
c. to place the activity under government control.
d. to create a public debate about the activity.

_____ **9.** The successful cleanup of the Nashua River can be credited to

a. an individual effort.
b. a political effort.
c. a business community effort.
d. All of the above

_____ **10.** The increased population of algae in Lake Washington was important because it signified which of the following?

a. The algae was becoming extinct.
b. Acid rain was changing the lake.
c. The lake was being polluted by fertilizers.
d. The ozone layer was thinner over Lake Washington.

Name ______________________ Class ______________ Date ______________

Assessment

Chapter Test

The Environment

In the space provided, write the letter of the term or phrase that best completes each statement or best answers each question.

______ **1.** The loss of species from some lakes in the northeastern United States may best be explained by
- **a.** global warming.
- **b.** evolutionary trends.
- **c.** the destruction of the ozone layer.
- **d.** acid rain.

______ **2.** The heat-trapping ability of some gases in the atmosphere is responsible for
- **a.** acid rain.
- **b.** the greenhouse effect.
- **c.** the creation of CFCs.
- **d.** increased levels of ultraviolet radiation.

______ **3.** Topsoil and ground water
- **a.** exist in unlimited quantities in aquifers throughout the world.
- **b.** are found only on the prairie.
- **c.** are renewable resources.
- **d.** are nonrenewable resources.

______ **4.** The human population increases in size when the
- **a.** death rate equals the birth rate.
- **b.** death rate exceeds the birth rate.
- **c.** birth rate equals the death rate.
- **d.** birth rate exceeds the death rate.

______ **5.** The increase in world population from 5 million people 10,000 years ago to 130 million people 8,000 years later was probably a result of
- **a.** a drastic reduction in the death rate at the time.
- **b.** more dependable food sources.
- **c.** trade routes between the continents.
- **d.** All of the above

______ **6.** Risk analysis involves
- **a.** supplying a population with information.
- **b.** predicting the consequences of environmental intervention.
- **c.** political action.
- **d.** education.

______ **7.** Human population growth is most rapid in
a. Europe.
b. the United States.
c. Japan.
d. developing countries.

______ **8.** Molecules of chemical pollutants become increasingly concentrated in higher trophic levels in a process called
a. biological accumulations.
b. toxic magnification.
c. biological magnification.
d. pollutant magnification.

Questions 9 and 10 refer to the figure below.

Amount of Carbon Dioxide in the Atmosphere

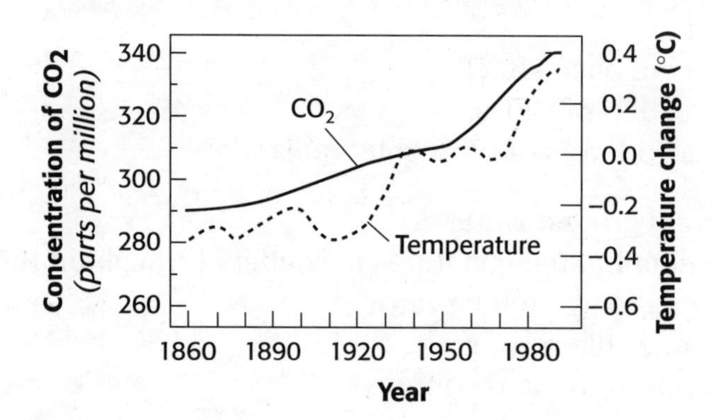

______ **9.** The graph above shows
a. the concentration of carbon dioxide in the atmosphere since 1860.
b. the average global temperature since 1860.
c. that the concentration of oxygen in the atmosphere has increased since 1860.
d. that the concentration of carbon dioxide in the atmosphere has decreased since 1860.

______ **10.** According to the graph above,
a. from 1900 to 1950, the average global temperature constantly increased.
b. the concentration of carbon dioxide in the atmosphere has increased at the same steady rate for the past 100 years.
c. the temperature and the concentration of carbon dioxide were the same in 1940.
d. the temperature has increased since 1980.

In the space provided, write the letter of the description that best matches the term or phrase.

______ **11.** malignant melanoma

______ **12.** scrubbers

______ **13.** knowledge of ecology

______ **14.** CFCs

______ **15.** aquifers

______ **16.** symbiotic fungi

______ **17.** DDT

______ **18.** blue-green algae

______ **19.** carbon dioxide

______ **20.** mercury

a. a measure taken that has effectively reduced pollution

b. insecticide banned in the United States

c. porous rock reservoirs that store ground water

d. pollutant responsible for wiping out fish in the Rhine River

e. a greenhouse gas

f. essential tool that allows an individual to solve environmental problems

g. chemicals that break down and destroy ozone

h. damaged by acid rain

i. increased incidences are a consequence of ozone depletion

j. increased presence can cause oxygen depletion in lakes

Name ______________________ Class ______________ Date ____________

Assessment

Chapter Test

The Environment

In the space provided, write the letter of the term or phrase that best completes each statement or best answers each question.

______ **1.** CFCs destroy ozone because ultraviolet radiation breaks the bonds in CFCs, and the
a. free chlorine atoms become more stable.
b. free chlorine atoms react to destroy ozone.
c. CFCs prevent ultraviolet radiation from reaching Earth's surface.
d. CFCs are released into the atmosphere.

______ **2.** Areas in the northeastern United States have been seriously affected by acid rain because
a. the northeastern United States produces a lot of sulfur pollution.
b. there is more precipitation in the northeastern United States.
c. dry air from the Southwest carries dust particles into the jet stream.
d. these areas are downwind from coal-burning plants in the Midwest.

______ **3.** Ozone depletion in the upper atmosphere can lead to increased incidences of which of the following?
a. skin cancer
b. asthma
c. cataracts
d. Both (a) and (c)

______ **4.** The precipitation in the northeastern United States is
a. about as acidic as the rest of the United States.
b. twice as acidic as the rest of the United States.
c. ten times as acidic as the rest of the United States.
d. one hundred times as acidic as the rest of the United States.

______ **5.** Although pesticides with chlorinated hydrocarbons are banned in the United States, they still pose an ecological threat because
a. they break down slowly in the environment.
b. they are still manufactured in the U.S. and they are used in other countries.
c. they become increasingly concentrated at the top of the food chain.
d. All of the above

______ **6.** What is the worst-case estimates of the world's species of plants and animals we could lose during the next 50 years?
a. 2 percent
b. 5 percent
c. 10 percent
d. 20 percent

______ **7.** Once pollution enters ground water,
- **a.** it is diluted into harmless trace amounts.
- **b.** it is costly to remove.
- **c.** there is no effective way to remove it.
- **d.** it can be isolated into a small sector of the aquifer.

______ **8.** Since 1950, the world has lost how much of its topsoil?
- **a.** one-fourth
- **b.** one-third
- **c.** one-half
- **d.** three-fourths

______ **9.** The difference between annual birth rate and annual death rate for the world is
- **a.** 9 deaths per 1,000 people.
- **b.** 30 births per 1,000 people.
- **c.** 21 births per 1,000 people.
- **d.** even, resulting in no net increase or decrease of people.

______ **10.** Providing dedicated lanes to cars with several occupants on highways causes
- **a.** a reduction in the number of cars on the road.
- **b.** gasoline prices to go down.
- **c.** increased traffic in all lanes.
- **d.** the development of more fuel-efficient engines.

In the space provided, write the letter of the description that best matches the term or phrase.

______ **11.** risk analysis

______ **12.** CFCs

______ **13.** lindane

______ **14.** scrubber

______ **15.** assessment

______ **16.** sulfur

- **a.** a chlorinated hydrocarbon compound used as a pesticide
- **b.** using information to predict the consequences of an environmental intervention
- **c.** the acidic element of acid rain
- **d.** chemicals used as heat exchangers that break down and destroy ozone
- **e.** a scientific analysis of an environmental problem
- **f.** reduces pollution emissions from industrial smokestacks

Read each question, and write your answer in the space provided.

17. What kinds of action can individuals take to help solve environmental problems?

____

____

____

18. What two approaches have been most effective in reducing pollution in the United States?

____

____

19. Explain the effects of biological magnification, using DDT as an example.

____

____

____

____

20. What five steps are followed to solve environmental problems successfully?

____

____

____

____

____

21. Explain why acid rain was not an environmental issue 200 years ago.

____

____

____

22. Why is it important to prevent the extinction of species?

____

____

____

23. How were chemical pollutants that washed into the Rhine River in Switzerland responsible for killing fish in the North Sea?

____

____

24. What do scientists know about the relationship between the greenhouse effect and global warming?

____

____

____

____

____

____

25. How is the population growth in developing countries different from the population growth in industrialized countries?

____

____

____

____

Name ______________________ Class ______________ Date ____________

Quick Lab

DATASHEET FOR IN-TEXT LAB

Modeling the Greenhouse Effect

You can use a quart jar to explore the greenhouse effect.

MATERIALS

- MBL or CBL system with appropriate software
- 2 temperature probes
- 1 qt jar
- lid with a 0.5 cm hole in the center
- tape
- heat source

Procedure

1. Set up an MBL/CBL system to collect data from two temperature probes at 6 second intervals for 150 data points.
2. Insert the end of one probe into the hole in the lid of a quart jar, and tape the probe in place. Place the other probe about 4 in. from the jar and at the same height as the first probe.
3. Place the jar about 30 cm from a heat-radiating source, and begin collecting data.
4. After 5 minutes, turn off (or remove) the heat source. Collect data for another 10 minutes.

Analysis

1. **Propose** an explanation for any differences between the two probes.

____

____

____

____

2. **Critical Thinking**
Comparing Functions How does carbon dioxide gas in the atmosphere function in a way similar to the glass jar?

____

____

____

3. Critical Thinking

Predicting Outcomes How would the temperature on Earth be different if there were no carbon dioxide in the atmosphere?

____

____

____

Name ______________________ Class ______________ Date ____________

Skills Practice Lab

DATASHEET FOR IN-TEXT LAB

Studying Population Growth

SKILLS

- Using a microscope
- Collecting, graphing, and analyzing data
- Calculating

OBJECTIVES

- **Observe** the growth and decline of a population of yeast cells.
- **Determine** the carrying capacity of a yeast culture.

MATERIALS

- safety goggles
- lab apron
- yeast culture
- (2) 1 mL pipets
- 2 test tubes
- 1% methylene blue solution
- ruled microscope slide (2 × 2 mm)
- coverslip
- compound microscope

CHEMSAFETY

CAUTION: Always wear safety goggles and a lab apron to protect your eyes and clothing.

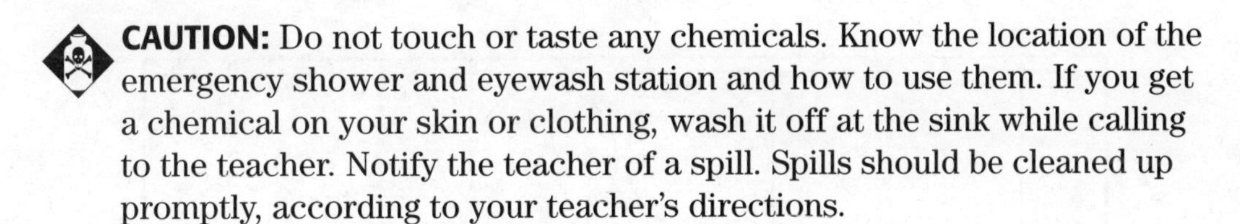

CAUTION: Do not touch or taste any chemicals. Know the location of the emergency shower and eyewash station and how to use them. If you get a chemical on your skin or clothing, wash it off at the sink while calling to the teacher. Notify the teacher of a spill. Spills should be cleaned up promptly, according to your teacher's directions.

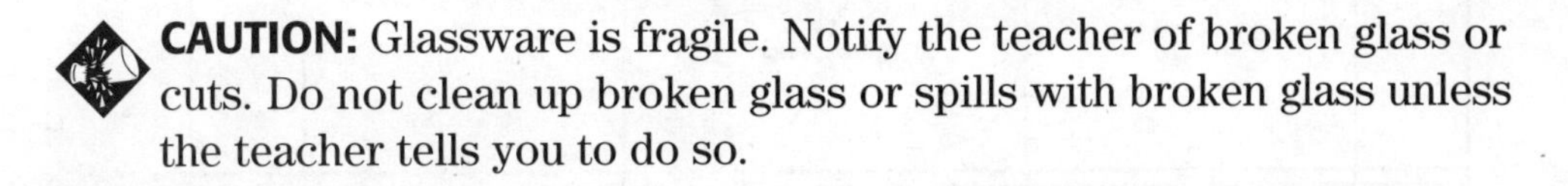

CAUTION: Glassware is fragile. Notify the teacher of broken glass or cuts. Do not clean up broken glass or spills with broken glass unless the teacher tells you to do so.

Before You Begin

Recall that population size is controlled by **limiting factors**—environmental resources such as food, water, oxygen, light, and living space. **Population growth** occurs when a population's **birth rate** is greater than its **death rate**. A decline in population size occurs when a population's death rate surpasses its

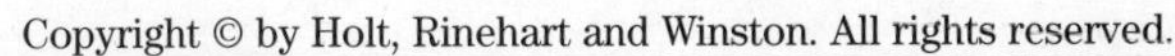

birth rate. In this lab, you will study the concepts of population growth, decline, and carrying capacity by growing and observing yeast.

1. Write a definition for each boldface term in the preceding paragraph. Use a separate sheet of paper.

2. You will be using the data table provided on the next page to record your data.

3. Based on the objectives for this lab, write a question about population growth that you would like to explore.

____

____

Procedure

PART A: COUNTING YEAST CELLS

1. Put on safety goggles and a lab apron.

2. Transfer 1 mL of a yeast culture to a test tube. Add 2 drops of methylene blue to the tube. **Caution: Methylene blue will stain your skin and clothing.** The methylene blue will remain blue in dead cells but will turn colorless in living cells.

3. Make a wet mount by placing 0.1 mL (one drop) of the yeast and methylene blue mixture on a ruled microscope slide. Cover the slide with a coverslip.

4. Observe the wet mount under the low power of a compound microscope. Notice the squares on the slide. Then switch to the high power. *Note: Adjust the light so that you can clearly see both stained and unstained cells.* Move the slide so that the top left-hand corner of one square is in the center of your field of view. This will be area 1, as shown in the diagram below.

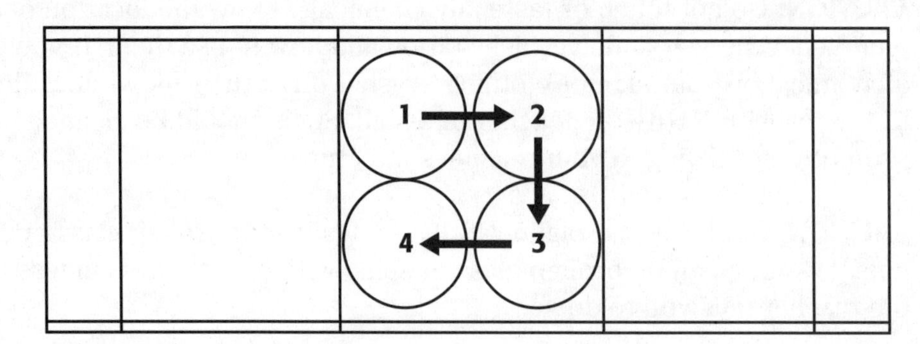

5. Count the live (unstained) cells and the dead (stained) cells in the four corners of a square using the pattern shown in the diagram. In the data table below, record the numbers of live cells and dead cells in the square.

Data Table			
Time (hours)	**Number of cells per square**		**Population size (cells/0.1 mL)**
	Squares 1–6	**Average**	
0			
24			
48			
72			
96			

6. Repeat step 5 until you have counted 6 squares on the slide. Complete Part B.

7. Find the total number of live cells in the 6 squares. Divide this total by 6 to find the average number of live cells per square. Record this number in the data table. Repeat this procedure for dead cells.

8. Estimate the population of live yeast cells in 1 mL (the amount in the test tube) by multiplying the average number of cells per square by 2,500. Record this number in the data table. Repeat this procedure for dead cells.

9. Repeat steps 1 through 8 each day for 4 more days.

PART B: CLEANUP AND DISPOSAL

10.  Dispose of solutions and broken glass in the designated waste containers. Do not pour chemicals down the drain or put lab materials in the trash unless your teacher tells you to do so.

11.  Clean up your work area and all lab equipment. Return lab equipment to its proper place. Wash your hands thoroughly before you leave the lab and after you finish all work.

Analyze and Conclude

1. **Analyzing Methods** Why were several areas and squares counted and then averaged each day?

____

____

____

2. **Summarizing Results** Use graph paper to graph the changes in the numbers of live yeast cells and dead yeast cells over time. Plot the number of cells in 1 mL of yeast culture on the y-axis and the time (in hours) on the x-axis.

3. **Inferring Conclusions** What limiting factors probably caused the yeast population to decline?

____

____

____

4. **Further Inquiry** Write a new question about population growth that could be explored in another investigation.

____

____

Name ______________________ Class ______________ Date ____________

Exploration Lab

CBL™ PROBEWARE

Effects of Acid Precipitation

When fossil fuels are burned, carbon dioxide, water vapor, and other gases are produced. Some of these gases react with water in the air to form acids, which eventually fall to Earth as precipitation. When this acid precipitation enters rivers, lakes, and other bodies of water, the pH of the water decreases.

Not all bodies of water are affected to the same extent, however. The pH of some bodies of water may change very little, while the pH of others may change a lot. The ability to resist changes in pH is known as *buffering capacity.* The buffering capacity of a body of water depends on the number and kinds of ions that are dissolved in the water. Natural waters that contain high concentrations of ions such as carbonate, CO_3^{2-}, and phosphate, PO_4^{3-}, are basic. These ions "buffer" the water by reacting with acids. As a result, the pH of the water does not change significantly when acid precipitation falls. Natural waters that contain fewer of these buffering ions are more susceptible to changes in pH.

Most of the ions that are dissolved in a lake's water are leached out of the rock that underlies the lake or carried into the lake by rivers and streams. Therefore, the composition of the rock in a region has an effect on the buffering capacity of the lakes in that region. In this lab, you will simulate the effects of acid precipitation on samples of water from different sources by adding acid to the samples and measuring their pH.

OBJECTIVES

Identify the effects of acid precipitation by adding drops of acid to several different water samples.

Test the pH of each water sample after adding each drop of acid.

Compare the buffering capacities of the water samples.

MATERIALS

- beaker, 500 mL
- beakers, 100 mL (4)
- buffer solution
- CBL System
- deionized water
- gloves
- H_2SO_4 solution, 0.10 M
- lab apron
- pH probe
- rinse bottle filled with deionized water
- safety goggles
- TI graphing calculator
- water samples (2)

Procedure

SETTING UP THE CBL SYSTEM

1. Plug the pH probe into the Channel 1 input of the CBL unit. Use the black cable to connect the CBL unit to the graphing calculator.
2. Turn on both the CBL unit and the calculator. Start the CHEMBIO program and go to the MAIN MENU.
3. Select SET UP PROBES. Enter "1" as the number of probes. Select pH from the SELECT PROBE menu. Enter "1" as the channel number.
4. Select USE STORED from the CALIBRATION menu.
5. Return to the MAIN MENU and select COLLECT DATA. Select MONITOR INPUT from the DATA COLLECTION menu. The CBL unit will display pH readings on the calculator.

COMPARING BUFFERING CAPACITIES

6. Put on safety goggles, gloves, and a lab apron. Pour 50 mL of deionized water into a 100 mL beaker.
7. Remove the pH probe from its storage solution. Use the rinse bottle filled with deionized water to carefully rinse the electrode, catching the rinse water in the 500 mL beaker.
8. Submerge the pH probe in the deionized water. When the pH reading stabilizes, record it in **Table 1.** This is the pH before any acid has been added.
9. Obtain a dropper bottle filled with dilute sulfuric acid, H_2SO_4. **CAUTION: Sulfuric acid is very corrosive. If you get sulfuric acid on your skin or clothing or in your eyes, rinse immediately with lukewarm water and notify your teacher. If you spill sulfuric acid on the floor or lab bench, notify your teacher. Avoid breathing the acid's fumes.**
10. Add a single drop of acid to the deionized water in the beaker. Swirl the solution gently, being careful not to hit the pH electrode with the side of the beaker. When the pH reading stabilizes, record it in **Table 1.**
11. Keep repeating step 10 until you have added 10 drops of acid to the beaker. Try to hold the dropper consistently so that the drops are all about the same size. Record the pH each time in **Table 1.**
12. Remove the pH electrode. Rinse it thoroughly with deionized water, catching the rinse water in the 500 mL beaker. Dispose of both the sample and the rinse water as instructed by your teacher.
13. Repeat steps 8–12 for the buffer solution and for each sample of water that you have. Use 50 mL of the solution or sample and a clean 100 mL beaker each time. Record the pH readings in **Table 1.**

14. When you finish, turn off both the CBL unit and the calculator. Rinse the pH electrode thoroughly with deionized water, and return it to its storage solution, making sure that the cap is on tight. Dispose of the sample and the rinse water as instructed by your teacher.

15. Put away your materials, and clean up your work area. Wash your hands thoroughly before leaving the lab.

TABLE 1 pH DATA

Drops of acid added	Deionized water	Buffer solution	Sample 1	Sample 2
0				
1				
2				
3				
4				
5				
6				
7				
8				
9				
10				

Analysis

1. Constructing Graphs Graph your data on the next page in **Figure 1.** Plot each sample of water that you tested, using a different color for each data set. Label your graph clearly for each of the four samples, or include a key.

FIGURE 1 EFFECT OF ACID ON pH OF WATER SAMPLES

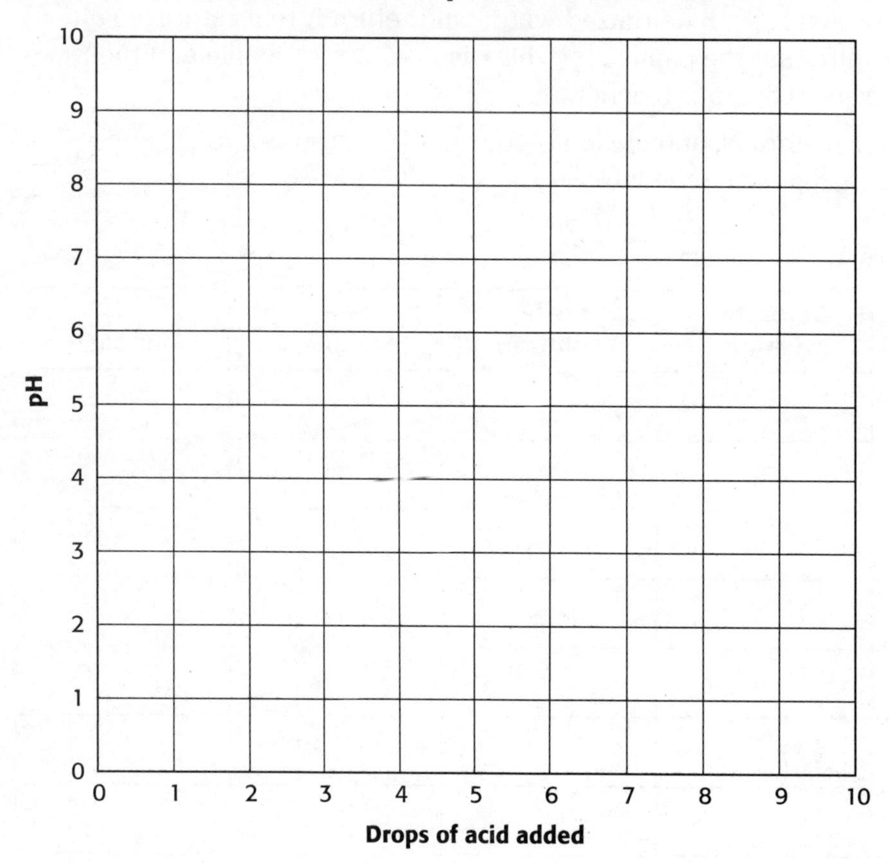

2. **Analyzing Graphs** Use your graph to compare the buffering capacities of deionized water and the buffer solution. Summarize the results of your comparison on the lines below.

____

____

____

____

3. **Analyzing Data** Which water sample has the higher buffering capacity? Use your data to support your answer.

____

____

____

____

Conclusions

1. **Interpreting Information** Why does deionized water have such a low buffering capacity?

____

____

____

2. **Interpreting Information** Explain why some of the water samples you tested have a greater buffering capacity than others.

____

____

3. **Interpreting Information** The pH scale is a logarithmic scale. This means that when the pH of a solution decreases by one unit, the hydrogen ion concentration increases by a factor of 10. Use this fact to explain why small changes in the pH of a body of water can be critical to the organisms that live there.

____

____

____

____

____

____

4. **Making Predictions** Reactions within your body are constantly generating acids, such as lactic acid and carbonic acid. You may also eat many foods that are high in citric acid, acetic acid, and other acids. However, the pH of your blood always stays between 7.35 and 7.45. Make a prediction about the buffering capacity of your blood. Which of the samples you tested is probably most like blood? Explain why.

____

____

____

____

____

5. Defending Conclusions Suppose someone suggests that one way to protect local lakes, streams, and rivers from acid precipitation is to add ions to the water to increase its buffering capacity. Do you think this is a good idea? Explain why or why not.

____

____

____

____

____

____

____

____

Extensions

1. Designing Experiments Develop a procedure to test the buffering capacities of several different antacids. Determine which of the antacids you tested has the greatest buffering capacity.

2. Research and Communications Collect data on the pH of a local body of water. Investigate the seasonal pattern of precipitation in your area and determine whether any variations in the pH of that body of water are correlated with precipitation. Present your findings in a written report.

Name ______________________ Class ______________ Date ____________

Inquiry Lab

CBL™ PROBEWARE

How Pollutants Affect a Lake

In this lab, you will design and conduct an experiment to determine how pollutants such as fertilizers and detergents affect the quality of water in a lake.

POSSIBLE MATERIALS

- "lake water" containing several different species of algae (100 mL)
- CBL system
- DO calibration bottle
- DO probe
- DO electrode filling solution
- fertilizer (nitrate) solution (10 mL)
- fluorescent lights or grow lamp
- lab apron
- laundry detergent (phosphate) solution (10 mL)
- link cable
- pH probe
- plastic graduated pipets (3)
- rinse bottle of deionized water
- safety goggles
- sheet of white paper
- small jars or 50 mL beakers (3)
- TI graphing calculator
- wax pencil

OBJECTIVES

Develop a hypothesis about how common pollutants affect the quality of lake water.

Design and **conduct** an experiment to test your hypothesis.

Identify relationships between common pollutants and the pH and DO content of lake water.

Evaluate your results.

Finding Out More Information

Lakes provide a home for a wide variety of organisms, including aquatic plants, fish, and a variety of arthropods, mollusks, and other invertebrates. The quality of the water in a lake affects the ability of these organisms to survive, grow, and reproduce. Aquatic organisms are sensitive to both the pH and the dissolved oxygen (DO) content of lake water. Organisms do best in lakes where the pH is between 6 and 9. A pH that is too high or too low can cause tissue damage and can increase the toxicity of compounds such as iron, ammonia, and mercury. Aquatic organisms are sensitive to the DO content of the lake water because they need oxygen to carry out cellular respiration. Cellular respiration provides these organisms with the energy they need to survive, grow, and reproduce.

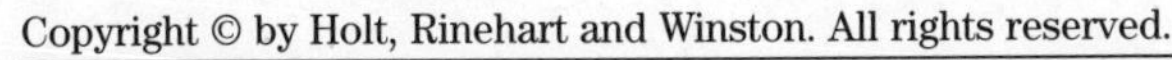

As rainwater runs off agricultural and residential lands, it often carries pollutants, such as fertilizers, detergents, and fecal material from farm animals, into lakes. Pollutants can have many effects on a lake. Some pollutants are toxins, some change the pH of the lake, and some are actually rich sources of nutrients. Nitrates and phosphates, which are present in fertilizers and laundry detergents, are nutrients that are beneficial for algae and plants in small amounts.

However, these nutrients become pollutants when they are present in excessive amounts. When excess nutrients are present, a sudden massive growth of algae called an *algal bloom* may result. The development of an algal bloom in a lake often causes the death of many aquatic plants and animals.

- What characteristics of lake water affect the health of aquatic organisms?

____

____

____

- How do pollutants such as nitrates and phosphates get into lake water?

____

____

____

____

- Aquatic organisms require nitrates and phosphates to live. Under what circumstances do they become pollutants?

____

____

Procedure

FORMING A HYPOTHESIS

Based on what you have learned, form a hypothesis about how fertilizers and detergents might create an unhealthy environment for aquatic organisms.

1. What characteristics of the lake water might be changed by the presence of excess nitrates and/or phosphates?

____

____

____

2. Write your own hypothesis. A possible hypothesis might be "The presence of excess nitrates changes the pH of the lake water to a level that is harmful to aquatic organisms."

____

____

COMING UP WITH A PLAN

Plan and conduct an experiment that will determine what changes the pollutants in the lake cause that might be harmful to the organisms living there. Limit the number of conditions you choose for your experiment to those that can be completed during the time your teacher has allotted for this lab. Consult with your teacher to make sure that the conditions you have chosen are appropriate.

3. Write out a procedure for your experiment on a separate sheet of paper. As you plan the procedure, make the following decisions.

- Decide what pollutant(s) you will use.
- Decide what characteristics of the "lake water" you will observe or measure.
- Select the materials and technology that you will need for your experiment from those that your teacher has provided.
- Decide where you will conduct your experiment.
- Decide what your control(s) will be.
- Decide what safety procedures are necessary.

4. Using graph paper or a computer, construct tables to organize your data. Be sure your tables fit your investigation.

5. Have your teacher approve your plans.

PERFORMING THE EXPERIMENT

6. Put on safety goggles and a lab apron.

7. Implement your plan, using the equipment, technology, and safety procedures that you selected. Instructions for using CBL probes to measure pH and dissolved oxygen are included on the next page.

8. Record your observations and measurements in your tables. If necessary, revise your tables to include variables that you did not think of while planning your experiment.

9. When you have finished, clean and store your equipment. Recycle or dispose of all materials as instructed by your teacher.

SETTING UP AND USING THE PH PROBE

10. Plug the pH probe into the Channel 1 input of the CBL unit. Use the black cable to connect the CBL unit to the graphing calculator.

11. Turn on both the CBL unit and the calculator. Start the CHEMBIO program and go to the MAIN MENU.

12. Select SET UP PROBES. Enter "1" as the number of probes. Select pH from the SELECT PROBE menu. Enter "1" as the channel number.

13. Select USE STORED from the CALIBRATION menu.

14. Return to the MAIN MENU and select COLLECT DATA. Select MONITOR INPUT for the DATA COLLECTION menu. The CBL unit will display pH readings on the calculator.

15. Remove the pH probe from its storage solution. Use the rinse bottle filled with deionized water to carefully rinse the probe, catching the rinse water in a 500 mL beaker.

16. Submerge the pH probe in your sample of "lake water." When the pH reading stabilizes, record the pH in your table. Rinse the pH probe with deionized water between each reading.

17. After the final reading, rinse the pH probe with deionized water and return the probe to its storage solution. Dispose of the rinse water as instructed by your teacher. Press "+" on the calculator.

SETTING UP AND USING THE DISSOLVED OXYGEN PROBE

18. Plug the dissolved oxygen probe into the Channel 1 input of the CBL unit. Use the black cable to connect the CBL unit to the graphing calculator.

19. Turn on both the CBL unit and the calculator. Start the CHEMBIO program and go to the MAIN MENU.

20. Select SET UP PROBES. Enter "1" as the number of probes. Select D.OXYGEN from the SELECT PROBE menu. Enter "1" as the channel number.

21. Select POLARIZE PROBE. Press ENTER to return to the CALIBRATION menu. You must allow the DO probe to polarize for 10 minutes before you can use it.

22. Select MANUAL ENTRY from the CALIBRATION menu. Enter the intercept (KO) and slope (K1) values for the DO calibration provided by your teacher.

23. After 10 minutes have passed, remove the DO probe from its storage solution. Submerge the probe in your sample of "lake water."

24. Select COLLECT DATA from the MAIN MENU. Select MONITOR INPUT from the DATA COLLECTION menu. Press ENTER.

25. Gently move the probe up and down about 1 cm in the sample. Be careful not to agitate the water, which will cause oxygen from the atmosphere to mix into the water. Continue moving the probe until the DO reading stabilizes. Record the DO concentration in your table.

26. Repeat steps 23 and 24 for each sample. Rinse the probe with deionized water between each reading.

27. After the final reading, rinse the DO probe with deionized water and return the probe to its storage solution. Press "+" on the calculator. Dispose of the rinse water as instructed by your teacher.

SETTING UP THE CBL SYSTEM FOR BOTH PROBES

To use both the pH probe and the DO probe, replace steps 12 and 20 with step 29, and replace steps 13 and 21 with steps 31 and 32.

28. Plug the pH probe into the Channel 1 input of the CBL unit. Plug the DO probe into the Channel 2 input. Use the black cable to connect the CBL unit to the graphing calculator.

29. Select SET UP PROBES. Enter "2" as the number of probes. Select pH from the SELECT PROBE menu. Enter "1" as the channel number.

30. Select MORE PROBES from the SELECT PROBE menu. Select D.OXYGEN from the SELECT PROBE menu. Enter "2" as the channel number.

31. Select POLARIZE PROBE. A message will appear. Select MANUAL ENTRY from the CALIBRATION menu. Enter the intercept (K0) and slope (K1) values for the dissolve oxygen calibration provided by your teacher. A message will appear concerning the sensors. Press ENTER. Leave the dissolved oxygen probe connected to the CBL for 10 minutes so that the probe can polarize.

32. Select COLLECT DATA from the MAIN MENU. Select MONITOR INPUT. Select either CH1 or CH2 from the SELECT A CHANNEL menu to monitor the probe reading. Use the CH VIEW button on the CBL to switch channels. Press TRIGGER on the CBL to quit monitoring. To view the other channel, select it from the SELECT A CHANNEL menu. To quit, choose QUIT from SELECT A CHANNEL menu.

Analysis

1. Summarizing Data Summarize your findings and observations.

____

____

____

____

____

____

____

2. **Describing Events** Share your results with your classmates. Which hypotheses were supported?

____

____

3. **Identifying Relationships** How might an algal bloom contribute to a decrease in dissolved oxygen in the "lake water"?

____

____

____

____

____

____

Conclusions

1. **Drawing Conclusions** What conclusions can you draw from your results? from class results?

____

____

____

2. **Evaluating Methods** Did your experimental design give clear results? If not, how might you improve your experimental design to give better results?

____

____

____

3. **Evaluating Models** Was your experiment a good model for how pollutants might affect lake water? Explain why or why not, and give examples of what might be missing from your model.

____

____

____

____

Name ______________________ Class ______________ Date ____________

Quick Lab

DATASHEET FOR IN-TEXT LAB

Modeling the Greenhouse Effect

You can use a quart jar to explore the greenhouse effect.

MATERIALS

- MBL or CBL system with appropriate software
- 2 temperature probes
- 1 qt jar
- lid with a 0.5 cm hole in the center
- tape
- heat source

Procedure

1. Set up an MBL/CBL system to collect data from two temperature probes at 6 second intervals for 150 data points.
2. Insert the end of one probe into the hole in the lid of a quart jar, and tape the probe in place. Place the other probe about 4 in. from the jar and at the same height as the first probe.
3. Place the jar about 30 cm from a heat-radiating source, and begin collecting data.
4. After 5 minutes, turn off (or remove) the heat source. Collect data for another 10 minutes.

Analysis

1. **Propose** an explanation for any differences between the two probes.

   The jar's glass kept heat from escaping, raising the temperature inside the jar.

2. **Critical Thinking**
   Comparing Functions How does carbon dioxide gas in the atmosphere function in a way similar to the glass jar?

   Like the glass of the jar, carbon dioxide retains heat energy from the sun that would otherwise dissipate.

Name ______________________ Class ______________ Date ____________

3. Critical Thinking

Predicting Outcomes How would the temperature on Earth be different if there were no carbon dioxide in the atmosphere?

The temperature of Earth would be much colder.

Name ______________________ Class ______________ Date ____________

Skills Practice Lab

DATASHEET FOR IN-TEXT LAB

Studying Population Growth

SKILLS

- Using a microscope
- Collecting, graphing, and analyzing data
- Calculating

OBJECTIVES

- **Observe** the growth and decline of a population of yeast cells.
- **Determine** the carrying capacity of a yeast culture.

MATERIALS

- safety goggles
- lab apron
- yeast culture
- (2) 1 mL pipets
- 2 test tubes
- 1% methylene blue solution
- ruled microscope slide (2 × 2 mm)
- coverslip
- compound microscope

CHEMSAFETY

CAUTION: Always wear safety goggles and a lab apron to protect your eyes and clothing.

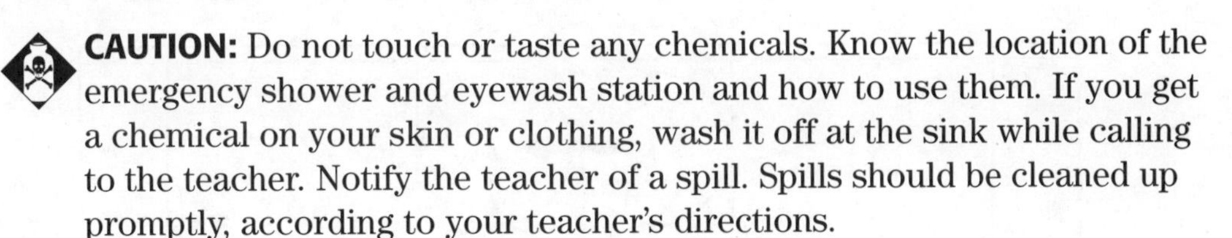

CAUTION: Do not touch or taste any chemicals. Know the location of the emergency shower and eyewash station and how to use them. If you get a chemical on your skin or clothing, wash it off at the sink while calling to the teacher. Notify the teacher of a spill. Spills should be cleaned up promptly, according to your teacher's directions.

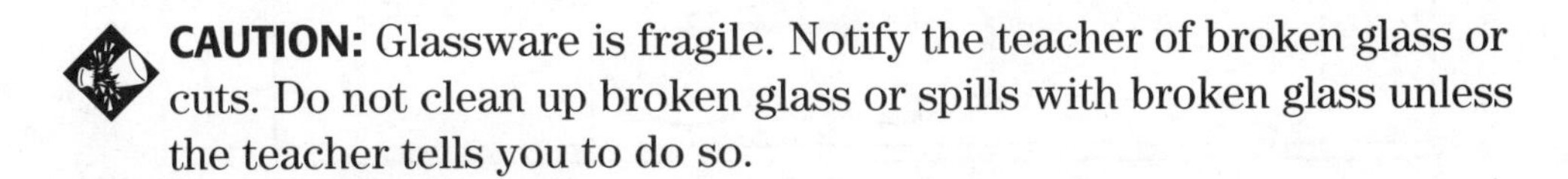

CAUTION: Glassware is fragile. Notify the teacher of broken glass or cuts. Do not clean up broken glass or spills with broken glass unless the teacher tells you to do so.

Before You Begin

Recall that population size is controlled by **limiting factors**—environmental resources such as food, water, oxygen, light, and living space. **Population growth** occurs when a population's **birth rate** is greater than its **death rate**. A decline in population size occurs when a population's death rate surpasses its

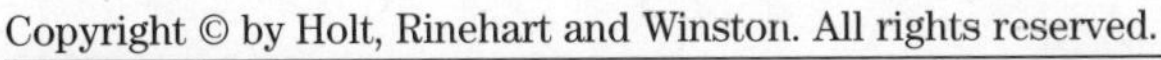

birth rate. In this lab, you will study the concepts of population growth, decline, and carrying capacity by growing and observing yeast.

1. Write a definition for each boldface term in the preceding paragraph. Use a separate sheet of paper. **Answers appear in the TE for this lab.**
2. You will be using the data table provided on the next page to record your data.
3. Based on the objectives for this lab, write a question about population growth that you would like to explore.

   For example: How does the size of a population with limited resources change over time?

Procedure

PART A: COUNTING YEAST CELLS

1. Put on safety goggles and a lab apron.
2. Transfer 1 mL of a yeast culture to a test tube. Add 2 drops of methylene blue to the tube. **Caution: Methylene blue will stain your skin and clothing.** The methylene blue will remain blue in dead cells but will turn colorless in living cells.
3. Make a wet mount by placing 0.1 mL (one drop) of the yeast and methylene blue mixture on a ruled microscope slide. Cover the slide with a coverslip.
4. Observe the wet mount under the low power of a compound microscope. Notice the squares on the slide. Then switch to the high power. *Note: Adjust the light so that you can clearly see both stained and unstained cells.* Move the slide so that the top left-hand corner of one square is in the center of your field of view. This will be area 1, as shown in the diagram below.

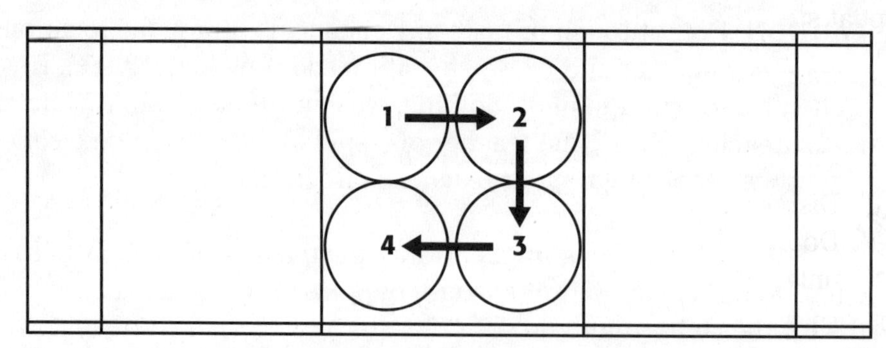

5. Count the live (unstained) cells and the dead (stained) cells in the four corners of a square using the pattern shown in the diagram. In the data table below, record the numbers of live cells and dead cells in the square.

Answers will vary.

Data Table			
Time (hours)	**Number of cells per square**		**Population size (cells/0.1 mL)**
	Squares 1–6	**Average**	
0			
24			
48			
72			
96			

6. Repeat step 5 until you have counted 6 squares on the slide. Complete Part B.

Answers will vary.

7. Find the total number of live cells in the 6 squares. Divide this total by 6 to find the average number of live cells per square. Record this number in the data table. Repeat this procedure for dead cells.

Answers will vary.

8. Estimate the population of live yeast cells in 1 mL (the amount in the test tube) by multiplying the average number of cells per square by 2,500. Record this number in the data table. Repeat this procedure for dead cells.

Answers will vary.

9. Repeat steps 1 through 8 each day for 4 more days.

Answers will vary.

PART B: CLEANUP AND DISPOSAL

10.  Dispose of solutions and broken glass in the designated waste containers. Do not pour chemicals down the drain or put lab materials in the trash unless your teacher tells you to do so.

11.  Clean up your work area and all lab equipment. Return lab equipment to its proper place. Wash your hands thoroughly before you leave the lab and after you finish all work.

Analyze and Conclude

1. **Analyzing Methods** Why were several areas and squares counted and then averaged each day?

   An average was taken to allow for variation within the population.

2. **Summarizing Results** Use graph paper to graph the changes in the numbers of live yeast cells and dead yeast cells over time. Plot the number of cells in 1 mL of yeast culture on the y-axis and the time (in hours) on the x-axis. **Answers appear in the TE for this lab.**

3. **Inferring Conclusions** What limiting factors probably caused the yeast population to decline?

   A lack of food and lack of space limit the yeast cells. They could also be poisoned by their own wastes.

4. **Further Inquiry** Write a new question about population growth that could be explored in another investigation.

   Answers will vary. For example: Would the carrying capacity of the yeast's environment expand if the size of the environment increased?

Exploration Lab

CBL™ PROBEWARE

Effects of Acid Precipitation

Teacher Notes

TIME REQUIRED One 45-minute period

SKILLS ACQUIRED

Collecting data
Experimenting
Identifying and recognizing patterns
Inferring
Interpreting
Measuring
Organizing and analyzing data
Predicting

RATINGS

Easy ← 1 2 3 4 → Hard

Teacher Prep–3
Student Setup–1
Concept Level–2
Cleanup–2

THE SCIENTIFIC METHOD

Make Observations Students observe the change in pH of deionized water, a buffer solution, and water samples as acid is added.

Analyze the Results Analysis questions 1–3 have students analyze their results.

Drawing Conclusions Students draw conclusions as they answer Conclusions questions 1, 2, 4, and 5.

MATERIALS

Materials for this lab can be purchased from WARD'S. See the *Master Materials List* for ordering instructions.

SAFETY CAUTIONS

- Discuss all safety symbols and caution statements with students.
- Emphasize to students that sulfuric acid is corrosive and can cause burns if it comes in contact with skin. Caution students also not to breathe the acid's fumes. If a chemical comes in contact with bare skin, have the student rinse the affected area immediately and continuously with lukewarm water. Review the Material Safety Data Sheet for sulfuric acid and for sodium hydroxide.

DISPOSAL

- Set out two disposal containers labeled "deionized water/acid waste" and "waste from water samples." Adjust the pH of the solutions in each container to between 5 and 9 with 0.10 M NaOH. Place each container in a sink separately, and run water to overflowing for 10 minutes, flushing to a sanitary sewer.

- The buffer solution waste can be washed down the drain with plenty of water, provided your school drains are connected to a sanitary sewer system.

TIPS AND TRICKS

Preparation

This lab works best in groups of two to four students.

You can either purchase 0.10 M H_2SO_4 or make it by adding 5.6 mL of concentrated sulfuric acid to a 1 L flask containing about 500 mL of deionized water and then filling to the 1 L mark with deionized water. **CAUTION: Wear safety goggles, gloves, and a lab apron. Be sure you add acid to water and not water to acid. Add the acid with extreme caution, as its addition to water can generate heat and explosive spattering.** Pour small quantities of the solution into individual dropper bottles for each group.

If possible, collect water samples from at least two local bodies of water. Refrigerate the samples if they must be stored, but allow them to come to room temperature before the experiment. Obtain water-quality data on the waters, if available, including alkalinity, total dissolved solids, hardness, and productivity.

If collecting local water samples is not practical, you can prepare samples that mimic waters with different buffering capacities by mixing deionized water with crushed limestone or clay (kaolin). Filter the mixtures and use the filtrates as water samples for students to test. A water sample that is a suitable substitute may be prepared by dissolving 2 g of sodium carbonate (Na_2CO_3) in 1 L of deionized water. For less or more buffering capacity, dissolve 1 g or 4 g of sodium carbonate per liter of deionized water, respectively.

CBL System

The procedure for this lab calls for using the stored calibration values for the pH probes. Recalibrating the pH probes might give more accurate results. To recalibrate, you can use two standard buffer solutions of pH 4 and pH 7. Refer to the pH probe booklet for instructions on calibrating probes.

You can either purchase pH 7 buffer solution or make it by combining 500 mL of 0.10 M Na_2PO_4 with 291 mL of 0.10 M NaOH and diluting to 1 L with deionized water. Check the pH and adjust to 7 if necessary.

Fresh deionized water should be used, as carbon dioxide will dissolve in the water over time and lower its pH.

The procedure in this lab is written for use with the original CBL system. If you are using CBL 2 or LabPro, the CHEMBIO program can still be used. Updated versions of this program can be downloaded from **www.vernier.com**. For additional information on how to integrate the CBL system into your laboratory, see the Program Introduction.

Some sensors may require the use of an adapter. Students will need to connect the adapter to the sensor before connecting it to the CBL.

Procedure

Tell students not to let their pH probe dry out. Caution them not to hit the sides or bottom of the beaker with the probe when they test the samples.

Remind students to rinse the probe with deionized water between samples to avoid contamination.

Name ______________________ Class ______________ Date ____________

Exploration Lab

CBL™ PROBEWARE

Effects of Acid Precipitation

When fossil fuels are burned, carbon dioxide, water vapor, and other gases are produced. Some of these gases react with water in the air to form acids, which eventually fall to Earth as precipitation. When this acid precipitation enters rivers, lakes, and other bodies of water, the pH of the water decreases.

Not all bodies of water are affected to the same extent, however. The pH of some bodies of water may change very little, while the pH of others may change a lot. The ability to resist changes in pH is known as *buffering capacity.* The buffering capacity of a body of water depends on the number and kinds of ions that are dissolved in the water. Natural waters that contain high concentrations of ions such as carbonate, CO_3^{2-}, and phosphate, PO_4^{3-}, are basic. These ions "buffer" the water by reacting with acids. As a result, the pH of the water does not change significantly when acid precipitation falls. Natural waters that contain fewer of these buffering ions are more susceptible to changes in pH.

Most of the ions that are dissolved in a lake's water are leached out of the rock that underlies the lake or carried into the lake by rivers and streams. Therefore, the composition of the rock in a region has an effect on the buffering capacity of the lakes in that region. In this lab, you will simulate the effects of acid precipitation on samples of water from different sources by adding acid to the samples and measuring their pH.

OBJECTIVES

Identify the effects of acid precipitation by adding drops of acid to several different water samples.

Test the pH of each water sample after adding each drop of acid.

Compare the buffering capacities of the water samples.

MATERIALS

- beaker, 500 mL
- beakers, 100 mL (4)
- buffer solution
- CBL System
- deionized water
- gloves
- H_2SO_4 solution, 0.10 M
- lab apron
- pH probe
- rinse bottle filled with deionized water
- safety goggles
- TI graphing calculator
- water samples (2)

Procedure

SETTING UP THE CBL SYSTEM

1. Plug the pH probe into the Channel 1 input of the CBL unit. Use the black cable to connect the CBL unit to the graphing calculator.
2. Turn on both the CBL unit and the calculator. Start the CHEMBIO program and go to the MAIN MENU.
3. Select SET UP PROBES. Enter “1” as the number of probes. Select pH from the SELECT PROBE menu. Enter “1” as the channel number.
4. Select USE STORED from the CALIBRATION menu.
5. Return to the MAIN MENU and select COLLECT DATA. Select MONITOR INPUT from the DATA COLLECTION menu. The CBL unit will display pH readings on the calculator.

COMPARING BUFFERING CAPACITIES

6. Put on safety goggles, gloves, and a lab apron. Pour 50 mL of deionized water into a 100 mL beaker.
7. Remove the pH probe from its storage solution. Use the rinse bottle filled with deionized water to carefully rinse the electrode, catching the rinse water in the 500 mL beaker.
8. Submerge the pH probe in the deionized water. When the pH reading stabilizes, record it in **Table 1.** This is the pH before any acid has been added.
9. Obtain a dropper bottle filled with dilute sulfuric acid, H_2SO_4. **CAUTION: Sulfuric acid is very corrosive. If you get sulfuric acid on your skin or clothing or in your eyes, rinse immediately with lukewarm water and notify your teacher. If you spill sulfuric acid on the floor or lab bench, notify your teacher. Avoid breathing the acid’s fumes.**
10. Add a single drop of acid to the deionized water in the beaker. Swirl the solution gently, being careful not to hit the pH electrode with the side of the beaker. When the pH reading stabilizes, record it in **Table 1.**
11. Keep repeating step 10 until you have added 10 drops of acid to the beaker. Try to hold the dropper consistently so that the drops are all about the same size. Record the pH each time in **Table 1.**
12. Remove the pH electrode. Rinse it thoroughly with deionized water, catching the rinse water in the 500 mL beaker. Dispose of both the sample and the rinse water as instructed by your teacher.
13. Repeat steps 8–12 for the buffer solution and for each sample of water that you have. Use 50 mL of the solution or sample and a clean 100 mL beaker each time. Record the pH readings in **Table 1.**

14. When you finish, turn off both the CBL unit and the calculator. Rinse the pH electrode thoroughly with deionized water, and return it to its storage solution, making sure that the cap is on tight. Dispose of the sample and the rinse water as instructed by your teacher.
15. Put away your materials, and clean up your work area. Wash your hands thoroughly before leaving the lab.

TABLE 1 pH DATA

Drops of acid added	Deionized water	Buffer solution	Sample 1	Sample 2
0	**6.81**	**7.01**	**7.42**	**7.45**
1	**5.46**	**6.99**	**7.31**	**7.29**
2	**4.04**	**6.99**	**7.25**	**6.98**
3	**3.51**	**6.94**	**7.18**	**6.75**
4	**3.43**	**6.94**	**7.00**	**6.51**
5	**3.38**	**6.98**	**6.89**	**6.25**
6	**3.31**	**6.96**	**6.81**	**5.96**
7	**3.26**	**6.95**	**6.75**	**5.69**
8	**3.23**	**6.87**	**6.71**	**5.46**
9	**3.18**	**6.83**	**6.66**	**5.22**
10	**3.15**	**6.81**	**6.63**	**5.03**

Entries will vary for each group. Sample data are entered above. Sample 1 is from a brackish-water lagoon. Sample 2 is from a productive golf course pond.

Analysis

1. **Constructing Graphs** Graph your data on the next page in **Figure 1.** Plot each sample of water that you tested, using a different color for each data set. Label your graph clearly for each of the four samples, or include a key.

Name ______________________ Class ______________ Date ____________

FIGURE 1 EFFECT OF ACID ON pH OF WATER SAMPLES

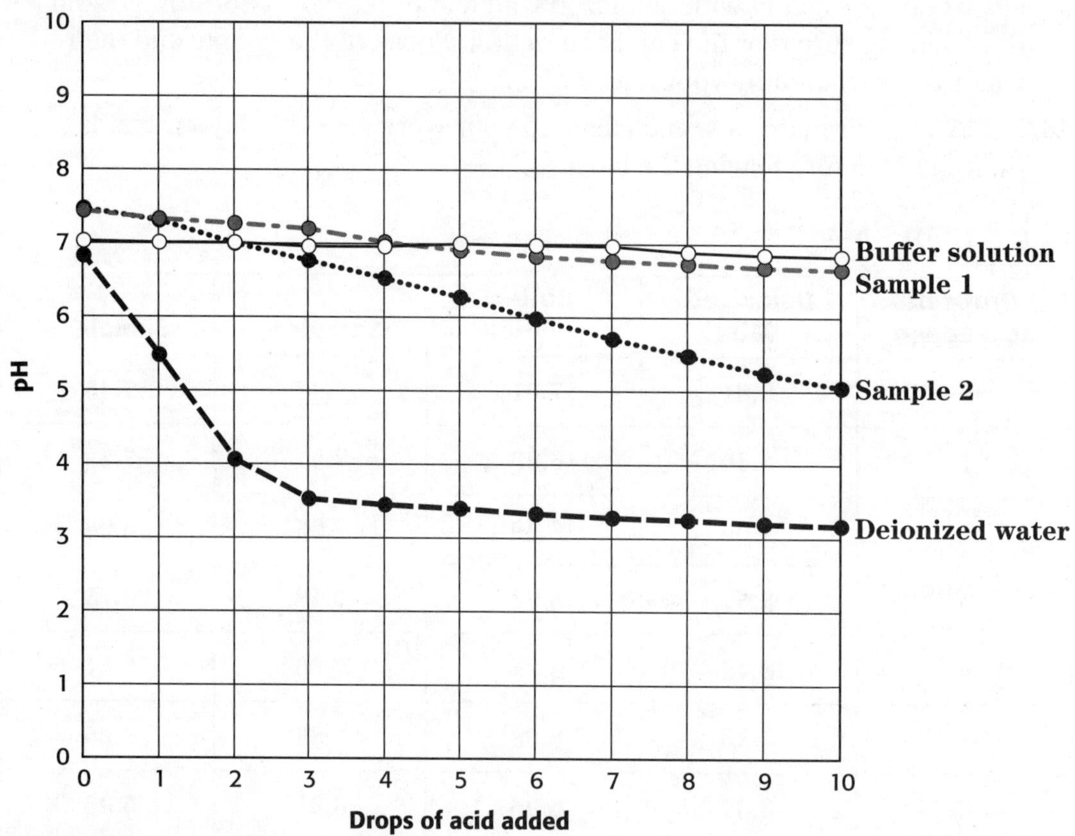

Graphs will vary, based on each group's data. The graph shown is based on the sample data in Table 1.

2. Analyzing Graphs Use your graph to compare the buffering capacities of deionized water and the buffer solution. Summarize the results of your comparison on the lines below.

The pH of deionized water dropped considerably when acid was added, so it must have very little buffering capacity. The pH of the buffer solution changed only slightly, so it must have a higher buffering capacity.

3. Analyzing Data Which water sample has the higher buffering capacity? Use your data to support your answer.

Answers will vary, but should be supported with data from the graph. For the sample data provided, the pH of sample 1 changed less than that of sample 2, so sample 1 has the higher buffering capacity.

Name ______________________ Class ______________ Date ____________

Conclusions

1. **Interpreting Information** Why does deionized water have such a low buffering capacity?

   Deionized water doesn't contain any dissolved salts that can react with and neutralize acids.

2. **Interpreting Information** Explain why some of the water samples you tested have a greater buffering capacity than others.

   Some waters have more dissolved salts that can react with and neutralize acids.

3. **Interpreting Information** The pH scale is a logarithmic scale. This means that when the pH of a solution decreases by one unit, the hydrogen ion concentration increases by a factor of 10. Use this fact to explain why small changes in the pH of a body of water can be critical to the organisms that live there.

   A small change in the pH of a body of water indicates a large change in the hydrogen ion concentration in the water. Hydrogen ions affect a variety of essential cellular functions, so changing their concentration beyond certain limits can cause death. Increasing the hydrogen ion concentration also causes the soil surrounding a body of water to release aluminum, which is toxic to some organisms.

4. **Making Predictions** Reactions within your body are constantly generating acids, such as lactic acid and carbonic acid. You may also eat many foods that are high in citric acid, acetic acid, and other acids. However, the pH of your blood always stays between 7.35 and 7.45. Make a prediction about the buffering capacity of your blood. Which of the samples you tested is probably most like blood? Explain why.

   Answers will vary. Students should predict that their blood has a high buffering capacity. For the sample data provided, the buffer solution and sample 1 are most like blood because their pH changed little when acid was added.

Name ______________________ Class ______________ Date ____________

5. Defending Conclusions Suppose someone suggests that one way to protect local lakes, streams, and rivers from acid precipitation is to add ions to the water to increase its buffering capacity. Do you think this is a good idea? Explain why or why not.

Answers will vary, but should be supported with a logical rationale. Students might suggest that adding ions to natural waters is a good idea because it would keep the pH of the waters from changing. Others might say that adding ions is a bad idea because the ions themselves might be toxic to organisms living in the waters or because it would take too many ions or be too expensive to treat all bodies of water in this way.

Extensions

1. **Designing Experiments** Develop a procedure to test the buffering capacities of several different antacids. Determine which of the antacids you tested has the greatest buffering capacity.
2. **Research and Communications** Collect data on the pH of a local body of water. Investigate the seasonal pattern of precipitation in your area and determine whether any variations in the pH of that body of water are correlated with precipitation. Present your findings in a written report.

Inquiry Lab

CBL™ PROBEWARE

How Pollutants Affect a Lake

Teacher Notes

TIME REQUIRED Two 45-minute periods, then 15 minutes every other day for five to seven days

SKILLS ACQUIRED

Collecting data
Experimenting
Identifying patterns
Inferring
Interpreting
Measuring
Organizing and analyzing data

RATINGS

Easy ← 1 2 3 4 → Hard

Teacher Prep–3
Student Setup–3
Concept Level–2
Cleanup–2

THE SCIENTIFIC METHOD

Make Observations Students make observations during their experiment.

Form a Hypothesis Procedure step 2 asks students to form a hypothesis.

Test the Hypothesis Procedure steps 3–5 guide students in designing an experiment that will test their hypothesis. Students conduct an experiment that tests their hypothesis.

Analyze the Results Students analyze results in Analysis questions 1 and 2.

Draw Conclusions Conclusions question 1 asks student to draw conclusions.

Communicate the Results Students communicate results in Analysis questions 1 and 2.

MATERIALS

Materials for this lab activity can be purchased from WARD'S. See the *Master Materials List* for ordering instructions.

Phosphate solution can be made with powdered laundry or dishwashing detergent or a powdered household cleaner (check label for phosphates and biodegradability) or with sodium phosphate, monobasic. Nitrate solution can be made with liquid or solid fertilizer containing no herbicides.

Prepare "lake water" by mixing any three or four of the following algae cultures: Spirogyra, Chlorella, Chamydomonas, Closetrium, Zygnema, Oscillatoria, Anabaena. Use tap water that has been left out to decholorinate for at least 24 hours.

DISPOSAL

Adjust the pH of the pollution solutions to neutrality and sterilize the algae with bleach before washing down the drain with copious amounts of water.

TECHNIQUES TO DEMONSTRATE

Demonstrate how to correctly handle probes (do not hit the bottom or sides of the beaker with the probe), and point out the fragile membrane at the tip of the dissolved oxygen (DO) probe.

TIPS AND TRICKS

This lab works best in groups of three to five students.

Prepare detergent/fertilizer solutions by adding 12.5 grams (e.g., powered detergent) or 12.5 mL (e.g., liquid fertilizer) and enough tap water to an Erlenmeyer flask to make 1 L. To break down the detergent and fertilizer so nutrients are available to the algae, add about 5 grams of dirt that is not sterile and that has no insecticides or herbicides added to it. Stopper the flask with cotton and leave it in a dark place for one week. Alternately, prepare phosphate solution by mixing 0.55 g sodium phosphate, monobasic with 99.45 mL distilled water. Filter the solutions before using them.

Refer to the DO probe booklet for instructions on polarizing and calibrating probes. The DO probes must be calibrated each day they are used. Provide students with the intercept and slope values you obtain. The DO probe must be polarized for 10 minutes. If more than one class is to perform this lab each day, students in earlier classes can leave their CBLs and calculators on when they finish. Students in the next class will then be able to skip all of the CBL setup, including the polarization.

This lab procedure calls for using the stored calibration values for pH probes. Recalibrating the pH probes might give more accurate results. To recalibrate, you can use two standard buffer solutions of pH 4 and pH 7.

CHECKPOINTS

1. By the end of the first class period, have students turn in a detailed one-page plan/procedure for approval.
2. During the second class period, have students revise procedures according to the teacher's approval and begin the procedure.
3. After the first data collection, have students turn in a copy of their data so that the teacher can ensure they are collecting data correctly.
4. Following completion of the experiment, have students turn in their procedure, data, and Analysis and Conclusions questions. Students' labs should be evaluated on lab technique, quality and clarity of observations, and the explanation of observations and conclusions.

SAMPLE PROCEDURE

Prepare three containers with 27 mL each of "lake water." Add 3 mL of fertilizer solution to one, 3 mL of phosphate solution to another, and 3 mL of water to the control. Place these preparations in an area where there is plenty of light for five to seven days. Use CBL probes to measure pH and/or dissolved oxygen on each day that class is held. Ideally, each group will test either the pH or dissolved oxygen of water containing one pollutant or the other. If different groups test different conditions, class data can be pooled to see the results for all conditions.

Name ______________________ Class ______________ Date ____________

Inquiry Lab

CBL™ PROBEWARE

How Pollutants Affect a Lake

In this lab, you will design and conduct an experiment to determine how pollutants such as fertilizers and detergents affect the quality of water in a lake.

POSSIBLE MATERIALS

- "lake water" containing several different species of algae (100 mL)
- CBL system
- DO calibration bottle
- DO probe
- DO electrode filling solution
- fertilizer (nitrate) solution (10 mL)
- fluorescent lights or grow lamp
- lab apron
- laundry detergent (phosphate) solution (10 mL)
- link cable
- pH probe
- plastic graduated pipets (3)
- rinse bottle of deionized water
- safety goggles
- sheet of white paper
- small jars or 50 mL beakers (3)
- TI graphing calculator
- wax pencil

OBJECTIVES

Develop a hypothesis about how common pollutants affect the quality of lake water.

Design and **conduct** an experiment to test your hypothesis.

Identify relationships between common pollutants and the pH and DO content of lake water.

Evaluate your results.

Finding Out More Information

Lakes provide a home for a wide variety of organisms, including aquatic plants, fish, and a variety of arthropods, mollusks, and other invertebrates. The quality of the water in a lake affects the ability of these organisms to survive, grow, and reproduce. Aquatic organisms are sensitive to both the pH and the dissolved oxygen (DO) content of lake water. Organisms do best in lakes where the pH is between 6 and 9. A pH that is too high or too low can cause tissue damage and can increase the toxicity of compounds such as iron, ammonia, and mercury. Aquatic organisms are sensitive to the DO content of the lake water because they need oxygen to carry out cellular respiration. Cellular respiration provides these organisms with the energy they need to survive, grow, and reproduce.

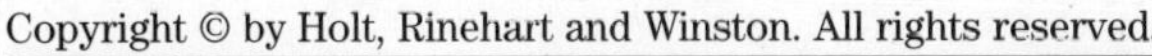

Name ______________________ Class ______________ Date ____________

How Pollutants Affect a Lake *continued*

As rainwater runs off agricultural and residential lands, it often carries pollutants, such as fertilizers, detergents, and fecal material from farm animals, into lakes. Pollutants can have many effects on a lake. Some pollutants are toxins, some change the pH of the lake, and some are actually rich sources of nutrients. Nitrates and phosphates, which are present in fertilizers and laundry detergents, are nutrients that are beneficial for algae and plants in small amounts.

However, these nutrients become pollutants when they are present in excessive amounts. When excess nutrients are present, a sudden massive growth of algae called an *algal bloom* may result. The development of an algal bloom in a lake often causes the death of many aquatic plants and animals.

- What characteristics of lake water affect the health of aquatic organisms?

  The pH of the lake and the amount of dissolved oxygen in the lake are characteristics that affect the health of aquatic organisms. Accept other reasonable answers.

- How do pollutants such as nitrates and phosphates get into lake water?

  Nitrates and phosphates are present in fertilizers, and phosphates are also present in household detergents. As rainwater runs off agricultural and residential land, it often becomes contaminated with these pollutants and carries them into lakes.

- Aquatic organisms require nitrates and phosphates to live. Under what circumstances do they become pollutants?

  Nitrates and phosphates become pollutants when they are present in excessive amounts. In excessive amounts, they cause the formation of an algal bloom.

Procedure

FORMING A HYPOTHESIS

Based on what you have learned, form a hypothesis about how fertilizers and detergents might create an unhealthy environment for aquatic organisms.

1. What characteristics of the lake water might be changed by the presence of excess nitrates and/or phosphates?

   Answers will vary. Students might suggest that excess nitrates and/or phosphates change any of the following: pH, DO content, amount of light available to organisms, or the amount of food available to organisms.

Name ______________________ Class ______________ Date ____________

2. Write your own hypothesis. A possible hypothesis might be "The presence of excess nitrates changes the pH of the lake water to a level that is harmful to aquatic organisms."

Answers will vary. Hypotheses should be testable in an experiment using materials from Possible Materials.

COMING UP WITH A PLAN

Plan and conduct an experiment that will determine what changes the pollutants in the lake cause that might be harmful to the organisms living there. Limit the number of conditions you choose for your experiment to those that can be completed during the time your teacher has allotted for this lab. Consult with your teacher to make sure that the conditions you have chosen are appropriate.

3. Write out a procedure for your experiment on a separate sheet of paper. As you plan the procedure, make the following decisions.
- Decide what pollutant(s) you will use.
- Decide what characteristics of the "lake water" you will observe or measure.
- Select the materials and technology that you will need for your experiment from those that your teacher has provided.
- Decide where you will conduct your experiment.
- Decide what your control(s) will be.
- Decide what safety procedures are necessary.

4. Using graph paper or a computer, construct tables to organize your data. Be sure your tables fit your investigation.

5. Have your teacher approve your plans.

PERFORMING THE EXPERIMENT

6. Put on safety goggles and a lab apron.

7. Implement your plan, using the equipment, technology, and safety procedures that you selected. Instructions for using CBL probes to measure pH and dissolved oxygen are included on the next page.

8. Record your observations and measurements in your tables. If necessary, revise your tables to include variables that you did not think of while planning your experiment.

9. When you have finished, clean and store your equipment. Recycle or dispose of all materials as instructed by your teacher.

Name ______________________ Class ______________ Date ____________

SETTING UP AND USING THE PH PROBE

10. Plug the pH probe into the Channel 1 input of the CBL unit. Use the black cable to connect the CBL unit to the graphing calculator.

11. Turn on both the CBL unit and the calculator. Start the CHEMBIO program and go to the MAIN MENU.

12. Select SET UP PROBES. Enter "1" as the number of probes. Select pH from the SELECT PROBE menu. Enter "1" as the channel number.

13. Select USE STORED from the CALIBRATION menu.

14. Return to the MAIN MENU and select COLLECT DATA. Select MONITOR INPUT for the DATA COLLECTION menu. The CBL unit will display pH readings on the calculator.

15. Remove the pH probe from its storage solution. Use the rinse bottle filled with deionized water to carefully rinse the probe, catching the rinse water in a 500 mL beaker.

16. Submerge the pH probe in your sample of "lake water." When the pH reading stabilizes, record the pH in your table. Rinse the pH probe with deionized water between each reading.

17. After the final reading, rinse the pH probe with deionized water and return the probe to its storage solution. Dispose of the rinse water as instructed by your teacher. Press "+" on the calculator.

SETTING UP AND USING THE DISSOLVED OXYGEN PROBE

18. Plug the dissolved oxygen probe into the Channel 1 input of the CBL unit. Use the black cable to connect the CBL unit to the graphing calculator.

19. Turn on both the CBL unit and the calculator. Start the CHEMBIO program and go to the MAIN MENU.

20. Select SET UP PROBES. Enter "1" as the number of probes. Select D.OXYGEN from the SELECT PROBE menu. Enter "1" as the channel number.

21. Select POLARIZE PROBE. Press ENTER to return to the CALIBRATION menu. You must allow the DO probe to polarize for 10 minutes before you can use it.

22. Select MANUAL ENTRY from the CALIBRATION menu. Enter the intercept (KO) and slope (K1) values for the DO calibration provided by your teacher.

23. After 10 minutes have passed, remove the DO probe from its storage solution. Submerge the probe in your sample of "lake water."

24. Select COLLECT DATA from the MAIN MENU. Select MONITOR INPUT from the DATA COLLECTION menu. Press ENTER.

25. Gently move the probe up and down about 1 cm in the sample. Be careful not to agitate the water, which will cause oxygen from the atmosphere to mix into the water. Continue moving the probe until the DO reading stabilizes. Record the DO concentration in your table.

26. Repeat steps 23 and 24 for each sample. Rinse the probe with deionized water between each reading.

27. After the final reading, rinse the DO probe with deionized water and return the probe to its storage solution. Press "+" on the calculator. Dispose of the rinse water as instructed by your teacher.

SETTING UP THE CBL SYSTEM FOR BOTH PROBES

To use both the pH probe and the DO probe, replace steps 12 and 20 with step 29, and replace steps 13 and 21 with steps 31 and 32.

28. Plug the pH probe into the Channel 1 input of the CBL unit. Plug the DO probe into the Channel 2 input. Use the black cable to connect the CBL unit to the graphing calculator.

29. Select SET UP PROBES. Enter "2" as the number of probes. Select pH from the SELECT PROBE menu. Enter "1" as the channel number.

30. Select MORE PROBES from the SELECT PROBE menu. Select D.OXYGEN from the SELECT PROBE menu. Enter "2" as the channel number.

31. Select POLARIZE PROBE. A message will appear. Select MANUAL ENTRY from the CALIBRATION menu. Enter the intercept (K0) and slope (K1) values for the dissolve oxygen calibration provided by your teacher. A message will appear concerning the sensors. Press ENTER. Leave the dissolved oxygen probe connected to the CBL for 10 minutes so that the probe can polarize.

32. Select COLLECT DATA from the MAIN MENU. Select MONITOR INPUT. Select either CH1 or CH2 from the SELECT A CHANNEL menu to monitor the probe reading. Use the CH VIEW button on the CBL to switch channels. Press TRIGGER on the CBL to quit monitoring. To view the other channel, select it from the SELECT A CHANNEL menu. To quit, choose QUIT from SELECT A CHANNEL menu.

Analysis

1. Summarizing Data Summarize your findings and observations.

Answers will depend on students' hypotheses and choices. The addition of nitrates to the "lake water" will likely cause an algal bloom. The DO content of the "lake water" decreases due to cellular respiration and decomposition of organisms in the algal bloom. The addition of nitrates and/or phosphates may change the pH of the water. The pH may decrease as the algal bloom releases carbon dioxide into the water, forming carbonic acid.

Name ______________________ Class ______________ Date ____________

2. Describing Events Share your results with your classmates. Which hypotheses were supported?

Answers will vary. See answers to Analysis question 1.

3. Identifying Relationships How might an algal bloom contribute to a decrease in dissolved oxygen in the "lake water"?

Answers will vary, but students should support their answers with logical rationales. Students might suggest the following: Algae use more oxygen during cellular respiration than they make during photosynthesis. The decomposition of dead algae uses up oxygen. The layer of algae prevents other photosynthetic organisms from getting enough light to carry out photosynthesis.

Conclusions

1. Drawing Conclusions What conclusions can you draw from your results? from class results?

Answers will vary, but students should recognize that the presence of the pollutants results in a decrease in water quality that may be harmful to many organisms.

2. Evaluating Methods Did your experimental design give clear results? If not, how might you improve your experimental design to give better results?

Answers will vary.

3. Evaluating Models Was your experiment a good model for how pollutants might affect lake water? Explain why or why not, and give examples of what might be missing from your model.

Answers will vary, but students should support their answers. Students might suggest that the effects of animals, plants, wind, rain, and soil are missing in their model.

Answer Key

Directed Reading

SECTION: GLOBAL CHANGE

1. b
2. d
3. a
4. c
5. d
6. c
7. b
8. a
9. b

SECTION: EFFECTS ON ECOSYSTEMS

1. Biological magnification refers to the way the concentration of harmful chemicals in the fatty tissue of animals increases as the chemicals move up through the trophic levels of the food chain.
2. Agricultural chemicals are chemicals such as pesticides, herbicides, and fertilizers that are introduced into the global ecosystem as a result of modern agriculture.
3. The tanker ran aground in 1989 off the coast of Alaska, and oil polluted about 1,600 km (1,000 mi) of coastline, causing extensive damage to wildlife.
4. Large amounts of industrial chemicals were released into the lake.
5. The presence of DDT in birds causes thin, fragile eggshells that can break during incubation.
6. Two potent anticancer drugs have been isolated from rosy periwinkle.
7. The species can only be found in Madagascar, which is a country that has been greatly damaged by deforestation.
8. Tropical rain forests contain a vast number of species of plants, and animals and are being destroyed at a very fast rate.
9. It takes tens of thousands of years for topsoil to accumulate, and topsoil cannot be replaced.
10. Aquifers are porous rock reservoirs for ground water.
11. Our ground water is being wasted through pollution and leaky faucets and through its use in watering lawns and washing cars.
12. d
13. d

SECTION: SOLVING ENVIRONMENTAL PROBLEMS

1. CFC
2. catalytic converters
3. scrubbers
4. tax
5. Collect data and perform experiments to construct a model of the ecosystem. The model is used to make predictions about the future course of the ecosystem.
6. Predict what could be expected to happen if a particular course of action was followed. Evaluate any adverse effects that a plan of action might have.
7. Explain the problem to the public, present the alternative actions available, and explain the probable costs and results of the different choices.
8. The public, through its elected officials, selects and implements a course of action.
9. The results of any action should be carefully monitored to see if the environmental problem is being solved.

Active Reading

SECTION: GLOBAL CHANGE

1. The cause is the burning of coal in power plants. The effect is the release of smoke containing high concentrations of sulfur.
2. to release the sulfur-rich smoke high into the atmosphere, where it could be dispersed and diluted by winds
3. It combines with water vapor to produce sulfuric acid.
4. Rain and snow carry the sulfuric acid to Earth's surface.
5. 3
6. 5
7. 2
8. 4
9. 1
10. b

SECTION: EFFECTS ON ECOSYSTEMS

1. The annual rate of 94 million is an estimated rather than an exact figure.

2. Answers will vary based on class time.
3. in the developing countries of Asia, Africa, and Latin America
4. in the industrialized countries of North America, Europe, Japan, New Zealand, and Australia
5. The rate in the United States of 0.8 percent is less than half the global rate. Doubling the U.S. rate yields 1.6 percent, so the world rate must be more than 1.6 percent.
6. a

SECTION: SOLVING ENVIRONMENTAL PROBLEMS

1. Data is collected and experiments are performed to determine exactly what is happening to the ecosystem.
2. A model makes it possible to describe how the ecosystem is responding to the problem and to predict future events in the ecosystem.
3. explaining the problem in understandable terms, presenting the alternative actions available, and explaining the probable costs and results of different choices
4. by exercising their right to vote and by contacting their elected officials
5. 6
6. 2
7. 7
8. 5
9. 1
10. 4
11. 3
12. d

Vocabulary Review

1. acid rain
2. chlorofluorocarbons or CFCs
3. greenhouse effect
4. biological magnification
5. aquifers
6. b
7. c
8. d
9. f
10. a
11. e

Science Skills

INTERPRETING MAPS

1. Sulfur emitted from coal-burning power plants combines with water vapor in the atmosphere to form sulfuric acid. When the water falls back to Earth as precipitation, it carries the sulfuric acid with it.
2. Chicago, IL; Louisville, KY; Birmingham, AL; Los Angeles, CA
3. The eastern United States receives the most acid rain. This region has more coal-burning power plants.
4. Coal-burning plants in the midwestern United States generate much of the chemical sources of acid rain. The sulfur is carried for hundreds of miles by the wind before returning to Earth in the form of acid rain.
5. the western United States

Concept Mapping

1. global change
2. ecosystems
3. acid rain or ozone depletion
4. ozone depletion or acid rain
5. pollution
6. agricultural chemicals
7. ground water or topsoil
8. topsoil or ground water

Critical Thinking

1. b
2. d
3. c
4. a
5. c
6. d
7. a
8. b
9. d
10. b
11. c
12. e
13. a
14. c, a
15. g, j
16. h, f
17. d, i
18. b, e
19. c
20. d
21. c
22. d

Test Prep Pretest

1. c
2. a
3. b
4. d
5. b

6. d
7. a
8. ultraviolet radiation
9. greenhouse effect
10. global warming
11. pesticides, herbicides, fertilizers
12. birth rate, death rate
13. growth rate
14. ground water
15. scrubbers
16. The ozone layer protects life on Earth from ultraviolet radiation.
17. With the greenhouse effect, gases such as carbon dioxide, methane, and nitrous oxide trap heat within the atmosphere. This trapped heat causes global temperatures to rise, an effect called global warming.
18. Molecules of polluting substances are transferred through successive levels of the food chain. As their molecules pass from one level to the next, polluting substances become more concentrated and their effects on complex organisms become more harmful.
19. The death rate has fallen steadily to an estimated level in 1994 of about 9 deaths per 1,000 people per year. Better sanitation and improved medical techniques have been primarily responsible for this drop.
20. assessment, risk analysis, public education, political action, and follow-through

Quiz

SECTION: GLOBAL CHANGE

1. f
2. d
3. a
4. e
5. c
6. b
7. b
8. d
9. d
10. b

SECTION: EFFECTS ON ECOSYSTEMS

1. b
2. d
3. e
4. a
5. c
6. c
7. c
8. d
9. a
10. d

SECTION: SOLVING ENVIRONMENTAL PROBLEMS

1. a
2. d
3. b
4. e
5. c
6. d
7. b
8. a
9. d
10. c

Chapter Test (General)

1. d
2. b
3. d
4. d
5. b
6. b
7. d
8. c
9. a
10. d
11. i
12. a
13. f
14. g
15. c
16. h
17. b
18. j
19. e
20. d

Chapter Test (Advanced)

1. b
2. d
3. d
4. c
5. d
6. d
7. c
8. b
9. c
10. a
11. b
12. d
13. a
14. f
15. e
16. c
17. Knowledge of ecology is an essential tool to solving environmental problems. Individuals can take steps to conserve energy such as by using bicycles or public transportation and by recycling or other conservation efforts.
18. passing laws that forbid pollution and placing taxes on pollution
19. Before its use was restricted, DDT was introduced liberally into the environment as an insecticide. The DDT passed from ground water and soil into plants and small animals and then to animals at higher trophic levels, becoming more concentrated at each level.

20. (1) Assess the problem by collecting data. (2) Perform a risk analysis to evaluate the various possible courses of action. (3) Educate the public about the most feasible course of action and its cost and expected results. (4) Implement a solution through political action. (5) Monitor actions in follow-through to verify that the problem is being effectively addressed and solved.

21. Acid rain began to be produced when smokestacks that released sulfur-rich smoke into the atmosphere were built. Such sources of sulfur smoke did not exist 200 years ago.

22. Many species offer possible benefits to humans, and we reduce our chances to learn about these species. Plant and animal species are used to develop improved food sources, medicine, or cures for diseases.

23. The chemicals were washed into the river in Switzerland. Then they flowed downstream through Germany and the Netherlands and out into the North Sea.

24. Since 1960, the levels of carbon dioxide in the atmosphere and the average global temperature have both risen steadily. The correlation of increasing temperatures with increasing carbon dioxide levels is very close. Many scientists believe that the two are related. However, correlation does not prove cause and effect. Both global temperature and levels of greenhouse gases may be changing because of other variables that have not yet been recognized.

25. Population growth in industrialized countries is slowing, with some industrialized countries even showing a decline in populations. Populations of most developing countries are increasing at rapid rates.

Lesson Plan

Section: Global Change

Pacing

Regular Schedule: **with lab(s):** N/A **without lab(s):** 3 days

Block Schedule: **with lab(s):** N/A **without lab(s):** 1 1/2 days

Objectives

1. Recognize the causes and effects of acid rain.
2. Evaluate the long-term consequences of atmospheric ozone.
3. Explain how the burning of fossil fuels has changed the atmosphere.
4. Recognize the relationship between the greenhouse effect and global warming.

National Science Education Standards Covered

UNIFYING CONCEPTS AND PROCESSES

UCP 1: Systems, order, and organization

UCP5: Form and function

SCIENCE AS INQUIRY

SI1: Abilities necessary to do scientific inquiry

SI2: Understandings about scientific inquiry

SCIENCE AND TECHNOLOGY

ST 1: Abilities of technological design

SCIENCE IN PERSONAL AND SOCIAL PERSPECTIVES

SPSP 2: Population growth

SPSP 3: Natural resources.

LIFE SCIENCE: INTERDEPENDENCE OF ORGANISMS

LSInter5: Human beings live within the world's ecosystems.

PHYSICAL SCIENCE

PS6: Interactions of energy and matter

KEY
SE = Student Edition
TE = Teacher Edition
CRF = Chapter Resource File

Block 1

CHAPTER OPENER *(45 minutes)*

_ **Quick Review,** SE. Students answer questions covered in previous sections of the textbook as preparation for the chapter content. **(GENERAL)**

_ **Reading Activity,** SE. Students write a short list of environmental issues they are familiar with and a list of questions about the evironment and environmental issues. **(GENERAL)**

_ **Using the Figure,** TE. Students answer questions about the chapter opener photograph. **(GENERAL)**

_ **Opening Activity**, To stimulate interest about the negative impact of humans on the environment, show students a picture that relays the magnitude of the problem. **(GENERAL)**

Block 2

FOCUS *(5 minutes)*

_ **Bellringer Transparency.** Use this transparency as students enter the classroom and find their seats. **(GENERAL)**

MOTIVATE *(10 minutes)*

_ **Discussion/Question**, TE. Students discuss the long-term effects of acid rain on the environment. **(GENERAL)**

TEACH *(30 minutes)*

_ **Teaching Transparency, Section Outline.** Use this transparency to give students a framework for the information in this section. **(GENERAL)**

_ **Integrating Physics and Chemistry**, TE. Students calculate the average pH of typical U.S. rainfall.

_ **Inclusion Strategies**, TE. Students design a postcard or travel poster of a forest and describe the effects of acid rain on the scene.

_ **Teaching Transparency, Ozone "Hole" Over Antarctica.** Use this transparency to discuss the thinning of the ozone layer over Antarctica. Point out that every since since 1978 ozone "hole" has grown larger. **(GENERAL)**

_ **Teaching Transparency, The Greenhouse Effect.** Use this transparency to compare Earth's greenhouse effect to heat being trapped in a greenhouse. **(GENERAL)**

HOMEWORK

_ **Active Reading Worksheet, Global Change, CRF.** Students read a passage related to the section topic and answer questions. **(GENERAL)**

_ **Directed Reading Worksheet, Global Change, CRF.** Students complete the exercises in this worksheet to help them understand the material as they read the section. **(BASIC)**

_ **Occupational Applications Worksheet, Wildlife Biologist, One-Stop Planner.** Students learn more about this job. **(GENERAL)**

Block 3

TEACH *(30 minutes)*

_ **Teaching Transparency, Amount of Carbon Dioxide in the Atmosphere.** Use this transparency to discuss the increase in carbon dioxide levels and average world temperatures. Inform students that of the carbon dioxide released by the burning of fossil fuel, about half remains in the atmosphere and half is absorbed by ocean waters. **(GENERAL)**

_ **Teaching Transparency, Changes in Global Temperature.** Use this transparency to point out the average change in global temperature since 1960. **(GENERAL)**

_ **Quick Lab,** Modeling the Greenhouse Effect, SE. Students study the greenhouse effect by measuring the temperature inside a covered jar and the ambient temperature near the jar. **(GENERAL)**

_ **Datasheets for In-Text Labs, Modeling the Greenhouse Effect, CRF.**

CLOSE *(15 minutes)*

_ **Alternative Assessment**, TE. Students write a short essay on ecosystem interdependence and hypothesize how damage to one ecosystem can effect other ecosystems. **(GENERAL)**

_ **Quiz,** TE. Students answer questions that review the section material. **(GENERAL)**

HOMEWORK

_ **Section Review,** SE. Assign questions 1–5 for review, homework, or quiz. **(GENERAL)**

_ **Quiz, CRF.** This quiz consists of ten multiple choice and matching questions that review the section's main concepts. **(BASIC) Also in Spanish.**

Other Resource Options

_ **Reteaching,** TE. Invite a guest speaker to address how your community might be affected by one of the following changes: acid rain, ozone depletion, or global warming. Have students prepare questions ahead of time. **(BASIC)**

- **Internet Connect.** Students can research Internet sources about Greenhouse Effect with SciLinks Code HX4094.
- **go.hrw.com.** For worksheets, videos, and other teaching aids related to this chapter, visit the HRW Web site and type in the keyword HX4 ENV.
- **CNN Science in the News, Video Segment 14 Greening Sudbury.** This video segment is accompanied by a **Critical Thinking Worksheet**.
- **CNN Science in the News, Video Segment 15 Tropical Reforestation.** This video segment is accompanied by a **Critical Thinking Worksheet**.
- **CNN Science in the News, Video Segment 24 Year of the Reef.** This video segment is accompanied by a **Critical Thinking Worksheet**.
- **Biology Interactive Tutor CD-ROM,** Unit 7 Ecosystem Dynamics. Students watch animations and other visuals as the tutor explains ecosystem dynamics. Students assess their learning with interactive activities.
- **CNN Student News.** Find the latest news, lesson plans, and activities related to important scientific events at **cnnstudentnews.com**.

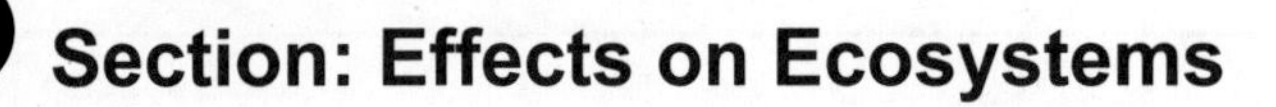

Lesson Plan

Section: Effects on Ecosystems

Pacing

Regular Schedule:	**with lab(s):** 5 days	**without lab(s):** 2 days
Block Schedule:	**with lab(s):** 2 1/2 days	**without lab(s):** 1 day

Objectives

1. Recognize the causes and effects of acid rain.
2. Evaluate the long-term consequences of atmospheric ozone.
3. Explain how the burning of fossil fuels has changed the atmosphere.
4. Recognize the relationship between the greenhouse effect and global warming.

National Science Education Standards Covered

SCIENCE AS INQUIRY

SI1: Abilities necessary to do scientific inquiry

SI2: Understandings about scientific inquiry

SCIENCE AND TECHNOLOGY

ST 1: Abilities of technological design

SCIENCE IN PERSONAL AND SOCIAL PERSPECTIVES

SPSP 2: Population growth

SPSP 3: Natural resources.

LIFE SCIENCE: THE CELL

LSCell1: Cells have particular structures that underlie their functions.

LSCell3: Cells store and use information to guide their functions.

LSCell4: Cell functions are regulated.

LIFE SCIENCE: INTERDEPENDENCE OF ORGANISMS

LSInter4: Human beings live within the world's ecosystems.

LSInter5: Human beings live within the world's ecosystems.

PHYSICAL SCIENCE

PS6: Interactions of energy and matter

KEY
SE = Student Edition
TE = Teacher Edition
CRF = Chapter Resource File

Block 4

FOCUS *(5 minutes)*

_ **Bellringer Transparency.** Use this transparency as students enter the classroom and find their seats. **(GENERAL)**

MOTIVATE *(10 minutes)*

_ **Activity**, TE. Students consider the causes and effects of different kinds of pollution shown in photos. Display pictures of different forms of environmental pollution, such as water pollution, air pollution, and solid waste and encourage students to add their own examples to the display.

TEACH *(30 minutes)*

_ **Teaching Transparency, Section Outline.** Use this transparency to give students a framework for the information in this section. **(GENERAL)**

_ **Teaching Transparency, Biological Magnification of DDT.** Use this transparency to discuss how DDT molecules pass up through the trophic food chain and become increasingly concentrated. **(GENERAL)**

_ **Teaching Transparency, Aquifer.** Use this transparency to discuss how the structure and function of aquifers. **(GENERAL)**

_ **Teaching Tip**, Ground-Water Contamination, TE. Students suggest different ways that chemicals might pollute ground water. **(GENERAL)**

HOMEWORK

_ **Teaching Tip**, Chemical Pollution, TE. Students research chemical pollution using the Web site in the Internet Connect information for this chapter. **(GENERAL)**

_ **Directed Reading Worksheet, Effects on Ecosystems, CRF.** Students complete the exercises in this worksheet to help them understand the material as they read the section. **(BASIC)**

_ **Active Reading Worksheet, Effects on Ecosystems, CRF.** Students read a passage related to the section topic and answer questions. **(GENERAL)**

Block 5

TEACH *(30 minutes)*

_ **Group Activity**, Ground Water in Your Back Yard, TE. Students research different ground water issues in their home city, region, or state. **(GENERAL)**

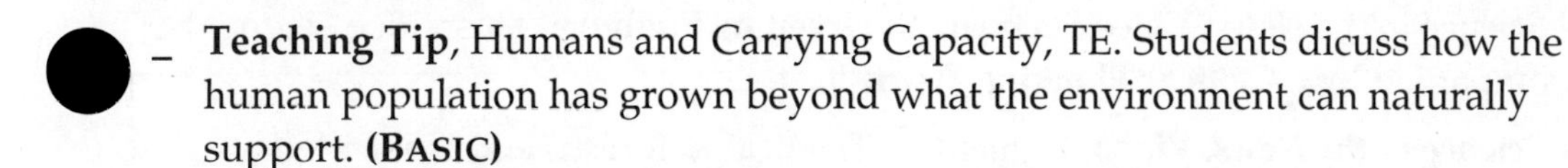

_ **Teaching Tip**, Humans and Carrying Capacity, TE. Students dicuss how the human population has grown beyond what the environment can naturally support. **(BASIC)**

_ **Teaching Transparency, World Population Growth Patterns.** Use this transparency to discuss projected human population growth. Review the current and projected populations for each country in the diagram. Contrast the rapid growth projected for developing countries with the slow growth predicted for industrialized countries. **(GENERAL)**

CLOSE *(15 minutes)*

_ **Quiz,** TE. Students answer questions that review the section material. **(GENERAL)**

_ **Reteaching,** TE. Students prepare an outline for a movie script about a real ecological disaster. **(BASIC)**

HOMEWORK

_ **Alternative Assessment**, TE. Students write an essay about which environmental concern will have their greatest impact on their community in the next decade. **(GENERAL)**

_ **Quiz, CRF.** This quiz consists of ten multiple choice and matching questions that review the section's main concepts. **(BASIC) Also in Spanish.**

_ **Section Review,** SE. Assign questions 1–5 for review, homework, or quiz. **(GENERAL)**

Optional Blocks

LABS *(135 minutes)*

_ **Exploration Lab, Effects of Acid Precipitation, CRF.** Students simulate the effects of acid precipitation on samples of water from different sources by adding acid to the samples and measuring their pH. **(GENERAL)**

_ **Inquiry Lab, How Pollutants Affect a Lake, CRF.** Students design and conduct an experiment to determine how pollutants such as fertilizers and detergents affect the quality of water in a lake. **(GENERAL)**

Other Resource Options

_ **Internet Connect.** Students can research Internet sources about Chemical Pollution with SciLinks Code HX4039.

_ **Internet Connect.** Students can research Internet sources about Aquifers with SciLinks Code HX4013.

_ **Internet Connect.** Students can research Internet sources about Population Growth with SciLinks Code HX4144.

_ **go.hrw.com.** For worksheets, videos, and other teaching aids related to this chapter, visit the HRW Web site and type in the keyword HX4 ENV.

_ **CNN Science in the News, Video Segment 14 Greening Sudbury.** This video segment is accompanied by a **Critical Thinking Worksheet**.

_ **CNN Science in the News, Video Segment 15 Tropical Reforestation.** This video segment is accompanied by a **Critical Thinking Worksheet**.

_ **CNN Science in the News, Video Segment 24 Year of the Reef.** This video segment is accompanied by a **Critical Thinking Worksheet**.

_ **Biology Interactive Tutor CD-ROM,** Unit 7 Ecosystem Dynamics. Students watch animations and other visuals as the tutor explains ecosystem dynamics. Students assess their learning with interactive activities.

_ **CNN Student News.** Find the latest news, lesson plans, and activities related to important scientific events at **cnnstudentnews.com**.

Lesson Plan

Section: Solving Environmental Problems

Pacing

Regular Schedule:	**with lab(s):** 6 days	**without lab(s):** 2 days
Block Schedule:	**with lab(s):** 5 days	**without lab(s):** 1 day

Objectives

1. Describe two effective approaches that have been taken to reduce pollution in the United States and abroad.
2. Evaluate the five major steps necessary to solve environmental problems.
3. Determine how individuals can take personal action to help solve environmental problems.

National Science Education Standards Covered

SCIENCE AS INQUIRY

SI1: Abilities necessary to do scientific inquiry

SI2: Understandings about scientific inquiry

SCIENCE AND TECHNOLOGY

ST 1: Abilities of technological design

SCIENCE IN PERSONAL AND SOCIAL PERSPECTIVES

SPSP 2: Population growth

SPSP 3: Natural resources.

LIFE SCIENCE: INTERDEPENDENCE OF ORGANISMS

LSInter4: Human beings live within the world's ecosystems.

LSInter5: Human beings live within the world's ecosystems.

PHYSICAL SCIENCE

PS6: Interactions of energy and matter

KEY
SE = Student Edition   **TE** = Teacher Edition
CRF = Chapter Resource File

Block 6

FOCUS *(5 minutes)*

_ **Bellringer Transparency.** Use this transparency as students enter the classroom and find their seats. **(GENERAL)**

MOTIVATE *(10 minutes)*

_ **Discussion/Question**, TE. Students discuss answers to this question: "Could an organism that does not have any impact on its environment exist?"

TEACH *(30 minutes)*

_ **Teaching Transparency, Section Outline.** Use this transparency to give students a framework for the information in this section. **(GENERAL)**

_ **Inclusion Strategies**, TE. Students design flyers title "Your Contribution" to distribute to classes in the school. The flyers should include examples and ideas of what students and their families can do to live more lightly on Earth.

_ **Demonstration**, TE. Show the class a catalytic converter or a cutaway diagram of one, and explain how it functions. **(GENERAL)**

_ **Real Life**, SE. Students research the arguments for and against recycling. **(GENERAL)**

HOMEWORK

_ **Directed Reading Worksheet, Solving Environmental Problems, CRF.** Students complete the exercises in this worksheet to help them understand the material as they read the section. **(BASIC)**

_ **Active Reading Worksheet, Solving Environmental Problems, CRF.** Students read a passage related to the section topic and answer questions. **(GENERAL)**

_ **Problem Solving Worksheet, Population Size, One-Stop Planner. (GENERAL)**

_ **Occupational Applications Worksheet, Blood-Bank Technologist, One-Stop Planner.** Students learn more about this job. **(GENERAL)**

Block 7

TEACH *(30 minutes)*

_ **Integrating Physics and Chemistry**, TE. Students identify the energy resource they are most dependent on and if there is a way to reduce their dependency.

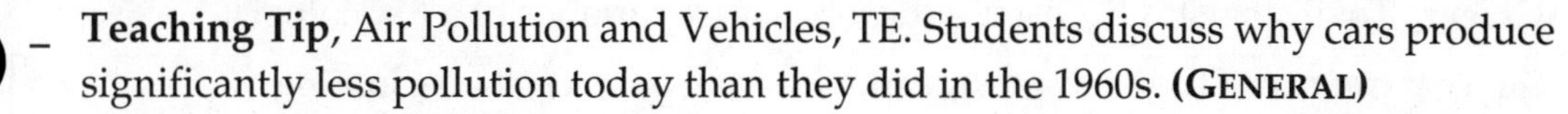

_ **Teaching Tip**, Air Pollution and Vehicles, TE. Students discuss why cars produce significantly less pollution today than they did in the 1960s. **(GENERAL)**

_ **Activity**, Technology, TE. Students use library resources to locate information about how an individual, group, or community solved an environmental problem using knowledge of biology. Students then make an oral presentation to the class. **(GENERAL)**

_ **Activity**, Making a Commitment, TE. Students consult books and magazines to identify some ways they personally can help the environment. Each student selects four actions that he or she is not already doing, carries out these actions for 2 weeks, and writes a brief an honest appraisal of his or her efforts. **(ADVANCED)**

CLOSE *(15 minutes)*

_ **Section Review,** SE. Assign questions 1–5. **(GENERAL)**

_ **Quiz,** TE. Students answer questions that review the section material. **(GENERAL)**

HOMEWORK

_ **Alternative Assessment,** TE. Students research the practice of selling pollution permits and write a report. **(GENERAL)**

_ **Science Skills Worksheet, CRF.** Students interpret a map showing the pH of rainfall in the United States. **(GENERAL)**

_ **Quiz, CRF.** This quiz consists of ten multiple choice and matching questions that review the section's main concepts. **(BASIC) Also in Spanish.**

_ **Modified Worksheet, One-Stop Planner.** This worksheet has been specially modified to reach struggling students. **(BASIC)**

_ **Critical Thinking Worksheet, CRF.** Students answer analogy-based questions that review the section's main concepts and vocabulary. **(ADVANCED)**

Optional Blocks

LAB *(20-30 minutes per day for 4 consecutive days)*

_ **Skills Practice Lab,** Studying Population Growth, SE. Students study the concepts of population growth, decline, and carrying capacity by growing and observing yeast for 4 consecutive days. **(GENERAL)**

_ **Datasheets for In-Text Labs, Studying Population Growth, CRF.**

Other Resource Options

_ **Reteaching,** TE. Have the class organize and implement a schoolwide project that addresses one of the environmental concerns discussed in this chapter. **(BASIC)**

_ **Supplemental Reading, Silent Spring, One-Stop Planner.** Students read the book and answer questions. **(ADVANCED)**

_ **Career,** Toxicologist, TE. Discuss the job and importance of a toxicologist.

_ **Internet Connect.** Students can research Internet sources about Solving Environmental Problems with SciLinks Code HX4166.

_ **go.hrw.com.** For worksheets, videos, and other teaching aids related to this chapter, visit the HRW Web site and type in the keyword HX4 ENV.

_ **CNN Science in the News, Video Segment 14 Greening Sudbury.** This video segment is accompanied by a **Critical Thinking Worksheet**.

_ **CNN Science in the News, Video Segment 15 Tropical Reforestation.** This video segment is accompanied by a **Critical Thinking Worksheet**.

_ **CNN Science in the News, Video Segment 24 Year of the Reef.** This video segment is accompanied by a **Critical Thinking Worksheet**.

_ **Biology Interactive Tutor CD-ROM,** Unit 7 Ecosystem Dynamics. Students watch animations and other visuals as the tutor explains ecosystem dynamics. Students assess their learning with interactive activities.

_ **CNN Student News.** Find the latest news, lesson plans, and activities related to important scientific events at **cnnstudentnews.com**.

Lesson Plan

End-of-Chapter Review and Assessment

Pacing

Regular Schedule: 2 days

Block Schedule: 1 day

KEY	
SE = Student Edition	TE = Teacher Edition
CRF – Chapter Resource File	

Block 8

REVIEW *(45 minutes)*

_ **Study Zone,** SE. Use the Study Zone to review the Key Concepts and Key Terms of the chapter and prepare students for the Performance Zone questions. **(GENERAL)**

_ **Performance Zone,** SE. Assign questions to review the material for this chapter. Use the assignment guide to customize review for sections covered. **(GENERAL)**

_ **Teaching Transparency, Concept Mapping.** Use this transparency to review the concept map for this chapter. **(GENERAL)**

Block 9

ASSESSMENT *(45 minutes)*

_ **Chapter Test, The Environment, CRF.** This test contains 20 multiple choice and matching questions keyed to the chapter's objectives. **(GENERAL) Also in Spanish.**

_ **Chapter Test, The Environment, CRF.** This test contains 25 questions of various formats, each keyed to the chapter's objectives. **(ADVANCED)**

_ **Modified Chapter Test, One-Stop Planner.** This test has been specially modified to reach struggling students. **(BASIC)**

Other Resource Options

_ **Vocabulary Review Worksheet, CRF.** Use this worksheet to review the chapter vocabulary. **(GENERAL) Also in Spanish.**

_ **Test Prep Pretest, CRF.** Use this pretest to review the main content of the chapter. Each question is keyed to a section objective. **(GENERAL) Also in Spanish.**

_ **Test Item Listing for ExamView® Test Generator, CRF.** Use the Test Item Listing to identify questions to use in a customized homework, quiz, or test.

_ **ExamView® Test Generator, One-Stop Planner.** Create a customized homework, quiz, or test using the HRW Test Generator program.

TEST ITEM LISTING

Human Impact on the Environment

TRUE/FALSE

1. ____ Acid rain usually has a pH between 8.3 and 9.7.
Answer: False Difficulty: I Section: 1 Objective: 1

2. ____ The phenomenon called acid rain helps many plants to grow stronger and taller.
Answer: False Difficulty: I Section: 1 Objective: 1

3. ____ Most countries have banned the use of CFCs, thus limiting the destruction of the ozone layer.
Answer: False Difficulty: I Section: 1 Objective: 2

4. ____ In the United States, the number of cases of malignant melanoma almost doubled since 1980.
Answer: True Difficulty: I Section: 1 Objective: 2

5. CFCs are the only chemicals that destroy ozone in the atmosphere.
Answer: False Difficulty: I Section: 1 Objective: 2

6. ____ The world's climate is warming as large amounts of carbon dioxide are released into the atmosphere.
Answer: True Difficulty: I Section: 1 Objective: 4

7. The environment is capable of absorbing any amount of pollution.
Answer: False Difficulty: I Section: 2 Objective: 1

8. Huge oil spills like the one in Alaska in 1989 account for the vast majority of pollution from oil seepage.
Answer: False Difficulty: I Section: 2 Objective: 1

9. If a lake is polluted with DDT, there will be a higher concentration of it in the water than in large fish in the lake.
Answer: False Difficulty: I Section: 2 Objective: 1

10. The presence of DDT in birds causes thin, fragile eggshells, most of which break during incubation.
Answer: True Difficulty: I Section: 2 Objective: 1

11. There is no benefit to using chemicals in agriculture.
Answer: False Difficulty: I Section: 2 Objective: 1

12. The use of chemicals in agriculture is necessary to meet the needs of an increasingly crowded world, but it must be done intelligently.
Answer: True Difficulty: I Section: 2 Objective: 1

13. Since use of DDT was banned in the United States, it has disappeared from use throughout the world.
Answer: False Difficulty: I Section: 2 Objective: 1

14. Companies in the United States continue to manufacture and export chlorinated hydrocarbons, even though they are banned here.
Answer: True Difficulty: I Section: 2 Objective: 1

15. Because of human efforts in the past decade, the preservation of nonrenewable resources such as topsoil, ground water, and our diverse species are no longer concerns for environmentalists.
Answer: False Difficulty: I Section: 2 Objective: 2

16. About one-half of Earth's tropical rain forests have been destroyed.
Answer: True Difficulty: I Section: 2 Objective: 2

17. The population of Earth has not changed very much over the last 350 years.
Answer: False Difficulty: I Section: 2 Objective: 3

18. The human birth rate has remained about the same for the last 350 years.
Answer: True Difficulty: I Section: 2 Objective: 3

19. The human population began to increase dramatically about 1500 years ago.
Answer: False Difficulty: I Section: 2 Objective: 3

20. The population of every country on Earth is steadily increasing.
Answer: False Difficulty: I Section: 2 Objective: 4

21. The exploding human population is the single greatest threat to the world's future.
Answer: True Difficulty: I Section: 2 Objective: 4

22. In the early 1990s there was a global increase in efforts to reduce pollution.
Answer: True Difficulty: I Section: 3 Objective: 1

23. There is no way known today to reduce emissions of sulfur dioxide, carbon monoxide, and soot from the air.
Answer: False Difficulty: I Section: 3 Objective: 1

24. Pollution and the economy of the United States are closely related.
Answer: True Difficulty: I Section: 3 Objective: 1

25. Political action is one of the five steps to successfully solve environmental problems.
Answer: True Difficulty: I Section: 3 Objective: 2

26. There is nothing that an individual can do to affect environmental problems because the problem is just too large and complex.
Answer: False Difficulty: I Section: 3 Objective: 3

27. W.T. Edmondson, at the University of Washington, proved that dumping sewage into Lake Washington was beneficial to the species diversity of the lake.
Answer: False Difficulty: I Section: 3 Objective: 3

28. As demonstrated in Lake Washington, sewage that has been treated and is safe enough to drink is not necessarily "harmless."
Answer: True Difficulty: I Section: 3 Objective: 3

MULTIPLE CHOICE

29. The decrease in species diversity of some lakes in the northeastern United States during the 20th century may best be explained by
a. global warming.
b. evolutionary trends.
c. the destruction of the ozone layer.
d. acid rain.
Answer: D Difficulty: I Section: 1 Objective: 1

30. Tall smokestacks were placed on power plants because the smoke they produced from the burning of coal contained high concentrations of
a. ozone.
b. sulfur.
c. oxygen.
d. nitrogen.
Answer: B Difficulty: I Section: 1 Objective: 1

31. A community that is *downwind* from a power plant that burns high-sulfur coal
   a. may experience the effects of acid rain.
   b. is safe since pollution from the plant is dispersed by the wind.
   c. will experience ozone depletion in the surrounding air.
   d. should pump extra oxygen into the air.

   Answer: A Difficulty: I Section: 1 Objective: 1

32. In the upper atmosphere, sulfur released from the burning of sulfur-rich coal combines with water vapor to form
   a. sulfur dioxide.
   b. sulfuric acid.
   c. ozone.
   d. All of the above

   Answer: B Difficulty: I Section: 1 Objective: 1

33. The destruction of the ozone layer may be responsible for an increase in
   a. cataracts.
   b. melanoma.
   c. cancer of the retina.
   d. All of the above

   Answer: D Difficulty: I Section: 1 Objective: 2

34. CFCs in the atmosphere
   a. result in free chlorine.
   b. change oxygen into ozone.
   c. convert sunlight into ozone.
   d. convert ozone into methane.

   Answer: A Difficulty: I Section: 1 Objective: 2

35. Chlorofluorocarbons (CFCs) are a problem because they
   a. corrode aerosol cans and release iron oxide into the atmosphere.
   b. are released by air conditioners into the ground water.
   c. attack ozone molecules in the upper atmosphere.
   d. were once thought to be a hazard, but this now causes unnecessary expense for industry.

   Answer: C Difficulty: I Section: 1 Objective: 2

36. CFCs were once
   a. thought to be chemically inert.
   b. used as refrigerants.
   c. used as aerosol propellants.
   d. All of the above

   Answer: D Difficulty: I Section: 1 Objective: 2

37. Ozone in the atmosphere
   a. leads to formation of acid precipitation.
   b. combines readily with water vapor.
   c. absorbs harmful radiation from the sun.
   d. All of the above

   Answer: C Difficulty: I Section: 1 Objective: 2

38. As a result of the discovery of the ozone hole,
   a. tall smokestacks were placed on power plants.
   b. the production of most CFCs was banned in the United States.
   c. methane has been substituted for nitrous oxides in some chemicals.
   d. large greenhouses were built in Europe, the United States, and Canada.

   Answer: B Difficulty: I Section: 1 Objective: 2

39. CFC : ozone ::
   a. ozone : carbon dioxide
   b. ozone : oxygen
   c. acid rain : fish and amphibians
   d. acid rain : carbon dioxide

   Answer: C Difficulty: II Section: 1 Objective: 2

40. The heat-trapping ability of some gases in the atmosphere can be compared to
    a. the melting of snow.
    b. the way glass traps heat in a greenhouse.
    c. condensation because of heating.
    d. heating water on a stove.

    Answer: B | Difficulty: I | Section: 1 | Objective: 4

41. Solar energy can be trapped in the atmosphere by
    a. the chemical bonds of carbon dioxide.
    b. ozone.
    c. radiation.
    d. sunlight.

    Answer: A | Difficulty: I | Section: 1 | Objective: 4

42. The greenhouse effect may increase on Earth because
    a. decomposers essential to recycling matter are being destroyed.
    b. too much oxygen is now given off by plants.
    c. increasing carbon dioxide will trap more heat.
    d. Earth tilts toward the sun in the summer.

    Answer: C | Difficulty: I | Section: 1 | Objective: 4

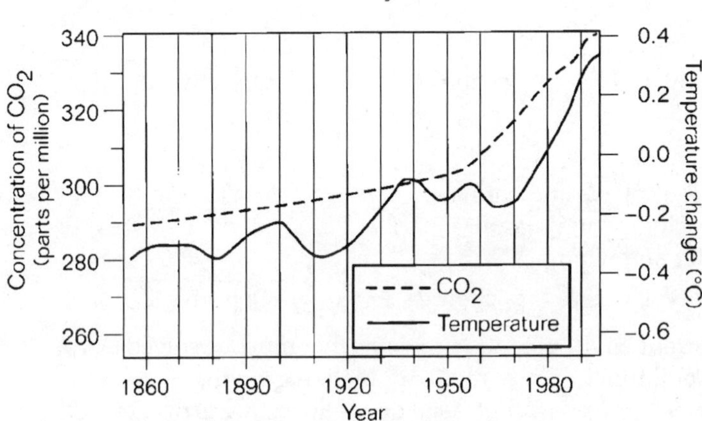

43. Refer to the illustration above. According to the graph,
    a. from 1900 to 1950, the average global temperature constantly increased.
    b. the concentration of CO_2 in the atmosphere increased at the same steady rate from 1920 to 1980.
    c. the concentration of CO_2 and the temperature were the same in 1940.
    d. None of the above

    Answer: D | Difficulty: II | Section: 1 | Objective: 4

44. Refer to the illustration above. The graph shows
    a. the concentration of CO_2 in the atmosphere from 1860 to beyond 1990.
    b. the average global temperature from 1860 to beyond 1990.
    c. that the concentration of oxygen in the atmosphere has increased since 1860.
    d. Both (a) and (b)

    Answer: D | Difficulty: II | Section: 1 | Objective: 4

45. If a large amount of chemicals gets into a river or lake, the species diversity
    a. is usually enhanced.
    b. recovers, but it takes years.
    c. recovers quickly.
    d. is usually unaffected.

    Answer: B | Difficulty: I | Section: 2 | Objective: 1

46. One way to help reduce pollution is to put a tax on the product contributing to the problem. To be fully effective, however, the tax must be
    a. high enough to reflect the actual cost of the pollution.
    b. very low so people will realize that they must donate money to help fight pollution.
    c. so high that no one can afford the product.
    d. the same throughout the world.

    Answer: A   Difficulty: I   Section: 2   Objective: 1

47. The extinction of species
    a. is a problem limited to the tropics.
    b. has been speeded up by the activities of people.
    c. is a problem only where topsoil and ground water are limited.
    d. is not a problem today.

    Answer: B   Difficulty: I   Section: 2   Objective: 2

48. Topsoil and ground water
    a. exist in unlimited quantities in aquifers throughout the world.
    b. are found only on the prairie.
    c. are renewable resources.
    d. are nonreplaceable resources.

    Answer: D   Difficulty: I   Section: 2   Objective: 2

49. Topsoil
    a. is a renewable resource.
    b. is formed from the remains of plants and animals.
    c. forms at the rate of 1 cm each growing season.
    d. forms by the action of water and wind.

    Answer: B   Difficulty: I   Section: 2   Objective: 2

50. All of the following are important environmental problems that must be solved *except*
    a. increasing levels of ocean pollution.
    b. dependence on fossil fuels.
    c. rapid population growth.
    d. coastal devastation by hurricanes.

    Answer: D   Difficulty: I   Section: 2   Objective: 2

51. Water trapped beneath the soil, largely in porous rock,
    a. is called ground water.
    b. is replenished immediately after a heavy rainstorm.
    c. is safe from pollution since it is deep beneath the soil.
    d. will never dry up.

    Answer: A   Difficulty: I   Section: 2   Objective: 2

52. Renewable sources of energy
    a. replenish themselves naturally.
    b. must be created in laboratories.
    c. are manufactured from fossil fuels.
    d. were never utilized until the 20th century.

    Answer: A   Difficulty: I   Section: 2   Objective: 2

53. Destruction of the tropical rain forests
    a. threatens the existence of thousands of species.
    b. provides for more pasture and farmlands.
    c. is done partly because of the need for lumber.
    d. All of the above

    Answer: D   Difficulty: I   Section: 2   Objective: 2

54. If a population is composed of a balance of people of pre-reproductive, reproductive, and post-reproductive age, what will most likely happen to the size of the population?
    a. It will grow steadily.
    b. It will experience no growth for a time and then increase rapidly.
    c. It will decrease steadily.
    d. It will experience no growth for a time and then decrease rapidly.

    Answer: A | Difficulty: I | Section: 2 | Objective: 3

55. A population of organisms grows
    a. when there are no natural restrictions except the availability of food.
    b. when the birth rate exceeds the death rate.
    c. only in the absence of predators or natural diseases.
    d. All of the above

    Answer: B | Difficulty: I | Section: 2 | Objective: 3

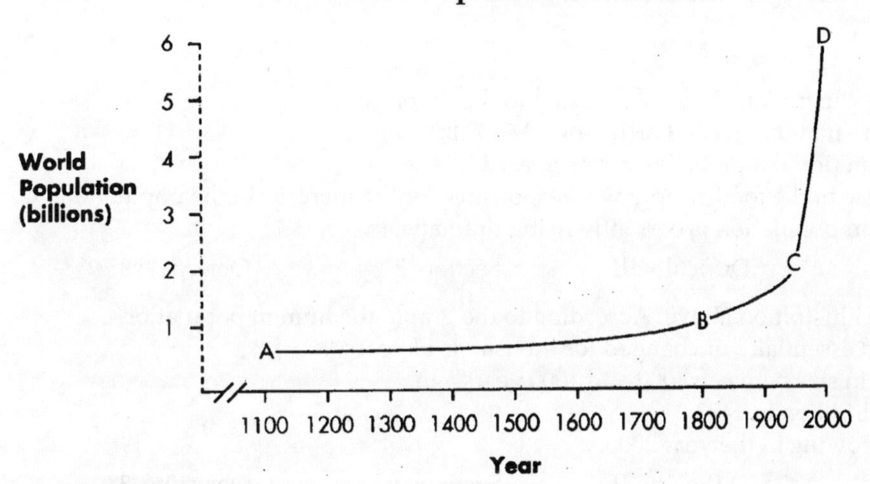

56. Refer to the illustration above. The American Revolution began in 1776. According to the graph, what was the approximate world population at that time?
    a. 500 thousand
    b. 1 million
    c. 1 billion
    d. 2 billion

    Answer: C | Difficulty: II | Section: 2 | Objective: 3

57. Refer to the illustration above. Which letter in the graph indicates the approximate world population in the year 1950?
    a. letter A
    b. letter B
    c. letter C
    d. letter D

    Answer: C | Difficulty: II | Section: 2 | Objective: 3

58. Refer to the illustration above. Which of the following contributed to the change in world population during the 1900s shown in the graph?
    a. better sanitation
    b. improved health care
    c. agricultural improvements
    d. All of the above

    Answer: D | Difficulty: II | Section: 2 | Objective: 3

59. Refer to the illustration above. The current rate of population growth will result in a doubling of the world population every 39 years. Based on information in the graph, what will be the approximate world population in the year 2039 if nothing is done to change this rate?
    a. 6 billion
    b. 10 billion
    c. 12 billion
    d. 24 billion

    Answer: C | Difficulty: II | Section: 2 | Objective: 3

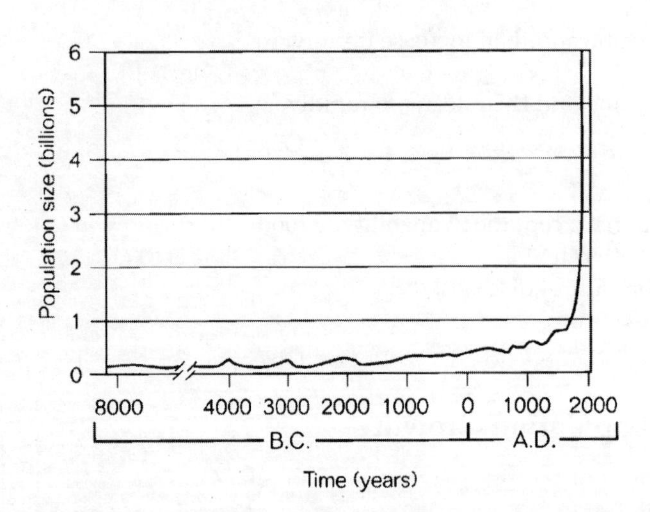

60. Refer to the illustration above. According to the graph,
    a. there were no humans on Earth around 6000 BC.
    b. the human population has never decreased in size.
    c. an increase in the food supply was responsible for the increase in the population.
    d. the human population is currently rising dramatically.

Answer: D　　Difficulty: II　　Section: 2　　Objective: 3

61. Refer to the illustration above. According to the graph, the human population
    a. remained essentially unchanged for thousands of years.
    b. doubled in size between 4000 and 1000 years ago.
    c. reached 1 billion in 1492.
    d. stoped growing in the year 2000.

Answer: A　　Difficulty: II　　Section: 2　　Objective: 3

62. Human population growth is most rapid in
    a. Europe.
    b. the United States.
    c. Japan.
    d. developing countries.

Answer: D　　Difficulty: I　　Section: 2　　Objective: 4

63. Pollutants produced by the burning of coal include
    a. chlorinated hydrocarbons.
    b. carbon dioxide.
    c. carbon monoxide.
    d. CFCs.

Answer: C　　Difficulty: I　　Section: 3　　Objective: 1

64. Which of the following can be effective in preventing pollution?
    a. stricter standards
    b. higher taxes on polluters
    c. better education of the public
    d. All of the above

Answer: D　　Difficulty: I　　Section: 3　　Objective: 1

65. What is the first step in solving an environmental problem?
    a. risk analysis
    b. political action
    c. assessment
    d. public education

Answer: C　　Difficulty: I　　Section: 3　　Objective: 2

66. The steps needed to solve environmental problems include
    a. waiting for the affected species to leave an ecosystem that is in trouble.
    b. leaving the problems to United Nations committees to address.
    c. educating the public about the problems and the costs of their solution.
    d. taking any necessary action, regardless of the consequences or adverse effects.

    Answer: C Difficulty: I Section: 3 Objective: 2

67. In solving environmental problems, which of the following is *not* a part of public education?
    a. explaining the problem in understandable terms
    b. sending out highly technical brochures that present detailed scientific research and complex data so everyone can become an expert on the problem
    c. presenting alternative actions available
    d. explaining the probable costs and results of the various choices

    Answer: B Difficulty: I Section: 3 Objective: 2

68. Things that you as an individual can do to contribute to a cleaner environment include
    a. recycling.
    b. using alternative means of transportation.
    c. helping educate the public.
    d. All of the above

    Answer: D Difficulty: I Section: 3 Objective: 3

COMPLETION

69. The formation of sulfuric acid in the atmosphere causes ____________________ ____________________.

    Answer: acid rain or acid precipitation Difficulty: II Section: 1 Objective: 1

70. The precipitation of sulfur-containing pollutants that lowers the natural pH of lakes and ponds, often killing the organisms that live there, is called ____________________ ____________________.

    Answer: acid rain Difficulty: II Section: 1 Objective: 1

71. Pollutants called ____________________ are converted into free chlorine that eventually destroys the protective ozone layer.

    Answer: CFCs or chlorofluorocarbons Difficulty: II Section: 1 Objective: 2

72. The heat-trapping ability of carbon dioxide, methane, and nitrous oxide in the atmosphere is known as the ____________________ ____________________.

    Answer: greenhouse effect Difficulty: II Section: 1 Objective: 3

73. Many scientists think that the increased levels of carbon dioxide in the atmosphere are linked to global ____________________.

    Answer: warming Difficulty: II Section: 1 Objective: 4

74. A knowledge of ____________________ is essential to solve our environmental problems.

    Answer: ecology Difficulty: II Section: 1 Objective: 4

75. As molecules of pollutants pass up through the trophic levels of the food chain, they become increasingly concentrated in a process called ____________________ ____________________.

    Answer: biological magnification Difficulty: II Section: 2 Objective: 1

76. Water trapped beneath the soil, largely in porous rock, is known as ____________________ ____________________.

    Answer: ground water Difficulty: II Section: 2 Objective: 2

77. The reason that ________________ resources must be conserved is that they cannot replenish themselves naturally.
Answer: nonrenewable Difficulty: I Section: 2 Objective: 2

78. A great deal of water is found beneath the soil within porous rock reservoirs called ________________.
Answer: aquifers Difficulty: II Section: 2 Objective: 2

79. The population of Earth is expected to reach ________________ billion people by the year 2025.
Answer: 8.5 Difficulty: II Section: 2 Objective: 3

80. The main reason that Earth's human population has increased over the past 350 years is because of a decrease in the ________________ rate.
Answer: death Difficulty: II Section: 2 Objective: 3

81. A population of organisms will grow when its ________________ ________________ exceeds its death rate.
Answer: birth rate Difficulty: II Section: 2 Objective: 3

82. The ________________ countries on Earth are experiencing the greatest increase in population growth.
Answer: developing Difficulty: II Section: 2 Objective: 4

83. The Clean Air Act of 1990 requires that power plants install scrubbers on their smokestacks to restrict ________________ emissions.
Answer: sulfur Difficulty: II Section: 3 Objective: 1

84. The economy of much of the industrialized world is based on a system of ________________ and demand.
Answer: supply Difficulty: II Section: 3 Objective: 1

85. In order to help solve environmental problems, one must be ________________ about the environment.
Answer: educated or knowledgeable Difficulty: II Section: 3 Objective: 2

86. The first step in addressing an environmental problem is ________________.
Answer: assessment Difficulty: II Section: 3 Objective: 2

87. The fifth and final step in successfully solving any environmental problem is ________________-________________.
Answer: follow, through Difficulty: II Section: 3 Objective: 2

88. In solving an environmental problem, predicting positive and negative consequences of environmental intervention is called ________________ ________________.
Answer: risk analysis Difficulty: II Section: 3 Objective: 2

ESSAY

89. Describe how acid precipitation in the form of rain or snow is formed.
Answer:
Sulfur in the smoke from burning sulfur-rich coal is released into the atmosphere. When condensation occurs, sulfuric acid forms from the combination of sulfur and water, and the sulfuric acid is brought back to Earth in the form of acid rain or snow.
Difficulty: II Section: 1 Objective: 1

90. John says that his teacher told him that ozone is a type of air pollutant produced by the burning of gasoline in automobiles. It has been determined to be harmful to humans. But he remembers reading that scientists and government officials are concerned about a decrease in the amount of Earth's ozone. Scientists are predicting that a decrease in the ozone layer will be harmful to humans. Write an explanation for John that will clarify the apparently conflicting pieces of information he has received.

    Answer:

    Ozone pollution from automobiles is released into the lower area of Earth's atmosphere, where it can be breathed in by humans. It can damage lung tissue. Earth's ozone layer is in the upper atmosphere. We do not breathe in this air. However, this higher level ozone layer acts to protect our health by absorbing ultra-violet radiation from the sun. With a depletion of Earth's ozone layer, more ultraviolet radiation would likely reach humans. Ultraviolet radiation causes mutations, some of which can result in cancer.

    Difficulty: III   Section: 1   Objective: 2

91. Describe the greenhouse effect.

    Answer:

    The greenhouse effect causes atmospheric temperatures to increase. The higher temperatures result from increased levels of carbon dioxide in the atmosphere. The increase in levels of carbon dioxide causes an increase in the ability of the atmosphere to trap heat, thus causing temperatures to rise gradually.

    Difficulty: II   Section: 1   Objective: 4

92. What is biological magnification?

    Answer:

    As molecules of pollutants pass up through the trophic levels of the food chain, they become increasingly concentrated.

    Difficulty: II   Section: 2   Objective: 1

93. Using examples, distinguish between renewable and nonrenewable natural energy resources.

    Answer:

    Renewable energy sources are readily and naturally replenished. For example, trees are a renewable resource, since new trees may be grown to replace those that are cut down. Nonrenewable resources are gone once they are used up; they cannot be replenished. Fossil fuels, such as coal, oil, and natural gas, are nonrenewable.

    Difficulty: II   Section: 2   Objective: 2

94. Why has our society made it profitable for industries to pollute the environment?

    Answer:

    Since indirect costs of pollution and environment damage are not included in the price that a consumer pays for a product, far more products are consumed than if indirect costs had been included. By not adding the indirect costs to the price of energy and manufactured goods, our society has made it profitable to pollute.

    Difficulty: III   Section: 3   Objective: 1

95. What are the five steps that must be followed in order to successfully solve an environmental problem?

Answer:

In order to solve the problem facing an ecosystem, it is necessary to

1. assess the problem by collecting data
2. perform a risk analysis to evaluate the various possible courses of action
3. educate the public about the most feasible course of action and its costs
4. implement a solution through political action
5. follow through on any action taken, to verify that the problem is being effectively addressed and solved.

Difficulty: II Section: 3 Objective: 2

97. Explain what was happening to Lake Washington in the 1950s because of the "harmless" sewage being dumped into it.

Answer:

The sewage was fertilizing the lake, thus providing an abundance of nutrients that allowed blue-green algae to grow in the water. The bacteria that was decomposing the dead algae was depleting the lake's oxygen and killing the lake.

Difficulty: II Section: 3 Objective: 3

9997265963 1 2 3 4 5 6